OFFICIAL
SQA
PAST
PAPERS
WITH ANSWERS

ADVANCED HIGHER

PHYSICS
2007-2011

✕SQA

iBright**RED**
PUBLISHING

Publisher's Note

We are delighted to bring you the 2011 Past Papers and you will see that we have changed the format from previous editions. As part of our environmental awareness strategy, we have attempted to make these new editions as sustainable as possible.

To do this, we have printed on white paper and bound the answer sections into the book. This not only allows us to use significantly less paper but we are also, for the first time, able to source all the materials from sustainable sources.

We hope you like the new editions and by purchasing this product, you are not only supporting an independent Scottish publishing company but you are also, in the International Year of Forests, not contributing to the destruction of the world's forests.

Thank you for your support and please see the following websites for more information to support the above statement –

www.fsc-uk.org

www.loveforests.com

© Scottish Qualifications Authority
All rights reserved. Copying prohibited. No part of this publication may be reproduced, stored in a retrieval system, or transmitted in any form or by any means, electronic, mechanical, photocopying, recording or otherwise.

First exam published in 2007.
Published by Bright Red Publishing Ltd, 6 Stafford Street, Edinburgh EH3 7AU
tel: 0131 220 5804 fax: 0131 220 6710 info@brightredpublishing.co.uk www.brightredpublishing.co.uk

ISBN 978-1-84948-232-5

A CIP Catalogue record for this book is available from the British Library.

Bright Red Publishing is grateful to the copyright holders, as credited on the final page of the Question Section, for permission to use their material. Every effort has been made to trace the copyright holders and to obtain their permission for the use of copyright material. Bright Red Publishing will be happy to receive information allowing us to rectify any error or omission in future editions.

ADVANCED HIGHER

2007

[BLANK PAGE]

X069/701

| NATIONAL QUALIFICATIONS 2007 | WEDNESDAY, 16 MAY 1.00 PM – 3.30 PM | PHYSICS ADVANCED HIGHER |

Reference may be made to the Physics Data Booklet.

Answer **all** questions.

Any necessary data may be found in the Data Sheet on page two.

Care should be taken to give an appropriate number of significant figures in the final answers to calculations.

Square-ruled paper (if used) should be placed inside the front cover of the answer book for return to the Scottish Qualifications Authority.

SCOTTISH QUALIFICATIONS AUTHORITY

DATA SHEET

COMMON PHYSICAL QUANTITIES

Quantity	Symbol	Value	Quantity	Symbol	Value
Gravitational acceleration on Earth	g	$9{\cdot}8 \text{ m s}^{-2}$	Mass of electron	m_e	$9{\cdot}11 \times 10^{-31}$ kg
Radius of Earth	R_E	$6{\cdot}4 \times 10^6$ m	Charge on electron	e	$-1{\cdot}60 \times 10^{-19}$ C
Mass of Earth	M_E	$6{\cdot}0 \times 10^{24}$ kg	Mass of neutron	m_n	$1{\cdot}675 \times 10^{-27}$ kg
Mass of Moon	M_M	$7{\cdot}3 \times 10^{22}$ kg	Mass of proton	m_p	$1{\cdot}673 \times 10^{-27}$ kg
Radius of Moon	R_M	$1{\cdot}7 \times 10^6$ m	Mass of alpha particle	m_α	$6{\cdot}645 \times 10^{-27}$ kg
Mean Radius of Moon Orbit		$3{\cdot}84 \times 10^8$ m	Charge on alpha particle		$3{\cdot}20 \times 10^{-19}$ C
Universal constant of gravitation	G	$6{\cdot}67 \times 10^{-11} \text{ m}^3 \text{ kg}^{-1} \text{ s}^{-2}$	Planck's constant	h	$6{\cdot}63 \times 10^{-34}$ J s
Speed of light in vacuum	c	$3{\cdot}0 \times 10^8 \text{ m s}^{-1}$	Permittivity of free space	ε_0	$8{\cdot}85 \times 10^{-12} \text{ F m}^{-1}$
Speed of sound in air	v	$3{\cdot}4 \times 10^2 \text{ m s}^{-1}$	Permeability of free space	μ_0	$4\pi \times 10^{-7} \text{ H m}^{-1}$

REFRACTIVE INDICES

The refractive indices refer to sodium light of wavelength 589 nm and to substances at a temperature of 273 K.

Substance	Refractive index	Substance	Refractive index
Diamond	2·42	Glycerol	1·47
Glass	1·51	Water	1·33
Ice	1·31	Air	1·00
Perspex	1·49	Magnesium Fluoride	1·38

SPECTRAL LINES

Element	Wavelength/nm	Colour	Element	Wavelength/nm	Colour
Hydrogen	656	Red	Cadmium	644	Red
	486	Blue-green		509	Green
	434	Blue-violet		480	Blue
	410	Violet		*Lasers*	
	397	Ultraviolet	Element	Wavelength/nm	Colour
	389	Ultraviolet	Carbon dioxide	9550 } 10590 }	Infrared
Sodium	589	Yellow	Helium-neon	633	Red

PROPERTIES OF SELECTED MATERIALS

Substance	Density/ kg m^{-3}	Melting Point/ K	Boiling Point/ K	Specific Heat Capacity/ J kg^{-1} K^{-1}	Specific Latent Heat of Fusion/ J kg^{-1}	Specific Latent Heat of Vaporisation/ J kg^{-1}
Aluminium	$2{\cdot}70 \times 10^3$	933	2623	$9{\cdot}02 \times 10^2$	$3{\cdot}95 \times 10^5$	. . . .
Copper	$8{\cdot}96 \times 10^3$	1357	2853	$3{\cdot}86 \times 10^2$	$2{\cdot}05 \times 10^5$	. . . .
Glass	$2{\cdot}60 \times 10^3$	1400	. . . .	$6{\cdot}70 \times 10^2$	. . . .	. . . .
Ice	$9{\cdot}20 \times 10^2$	273	. . . .	$2{\cdot}10 \times 10^3$	$3{\cdot}34 \times 10^5$	. . . .
Glycerol	$1{\cdot}26 \times 10^3$	291	563	$2{\cdot}43 \times 10^3$	$1{\cdot}81 \times 10^5$	$8{\cdot}30 \times 10^5$
Methanol	$7{\cdot}91 \times 10^2$	175	338	$2{\cdot}52 \times 10^3$	$9{\cdot}9 \times 10^4$	$1{\cdot}12 \times 10^6$
Sea Water	$1{\cdot}02 \times 10^3$	264	377	$3{\cdot}93 \times 10^3$	. . . .	. . . .
Water	$1{\cdot}00 \times 10^3$	273	373	$4{\cdot}19 \times 10^3$	$3{\cdot}34 \times 10^5$	$2{\cdot}26 \times 10^6$
Air	1·29	. . . .	. . . .	. . . .	. . . .	. . . .
Hydrogen	$9{\cdot}0 \times 10^{-2}$	14	20	$1{\cdot}43 \times 10^4$	. . . .	$4{\cdot}50 \times 10^5$
Nitrogen	1·25	63	77	$1{\cdot}04 \times 10^3$	. . . .	$2{\cdot}00 \times 10^5$
Oxygen	1·43	55	90	$9{\cdot}18 \times 10^2$	. . . .	$2{\cdot}40 \times 10^5$

The gas densities refer to a temperature of 273 K and a pressure of $1{\cdot}01 \times 10^5$ Pa.

Marks

1. (a) A particle has displacement $s = 0$ at time $t = 0$ and moves with constant acceleration a.

   The velocity of the object is given by the equation $v = u + at$, where the symbols have their usual meanings.

   Using calculus, derive an equation for the displacement s of the object as a function of time t. 　　2

   (b) A cyclotron accelerates protons to a velocity of $2 \cdot 80 \times 10^8 \, \mathrm{m\,s^{-1}}$.

   Calculate the relativistic energy of a proton at this velocity. 　　4

   　　　　　　　　　　　　　　　　　　　　　　　　　　　　　　　(6)

[Turn over

Marks

2. (a) A turntable consists of a uniform disc of radius 0·15 m and mass 0·60 kg.

    (i) Calculate the moment of inertia of the turntable about the axis of rotation shown in Figure 1. **2**

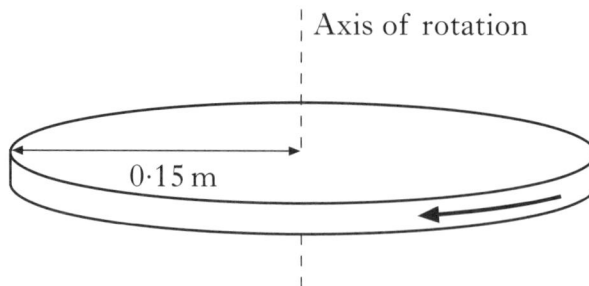

Figure 1

    (ii) The turntable accelerates uniformly from rest until it rotates at 45 revolutions per minute. The time taken for the acceleration is 1·5 s.

        (A) Show that the angular velocity after 1·5 s is 4·7 rad s^{-1}. **1**

        (B) Calculate the angular acceleration of the turntable. **2**

    (iii) When the turntable is rotating at 45 revolutions per minute, its motor is disengaged. The turntable continues to rotate freely with negligible friction.

    A small mass of 0·20 kg is dropped onto the turntable at a distance of 0·10 m from the centre, as shown in Figure 2.

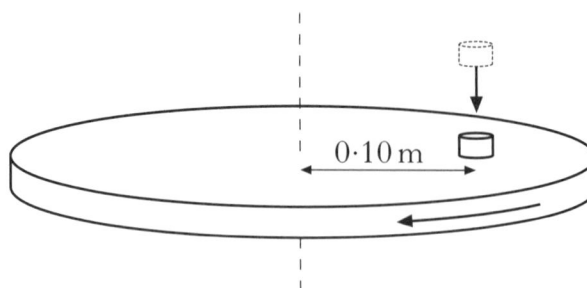

Figure 2

    The mass remains in this position on the turntable due to friction, and the turntable and mass rotate together.

    Calculate the new angular velocity of the turntable and mass. **3**

Marks

2. (a) (continued)

(iv) The experiment is repeated, but the mass is dropped at a distance greater than 0·10 m from the centre of the turntable. The mass slides off the turntable.

Explain why this happens.

2

(b) An ice-skater spins with her arms and one leg outstretched as shown in Figure 3(a). She then pulls her arms and leg close to her body as shown in Figure 3(b).

Figure 3(a)

Figure 3(b)

State what happens to her angular velocity during this manoeuvre.

Justify your answer.

2

(12)

[Turn over

Marks

3. (*a*) The Moon orbits the Earth due to the gravitational force between them.

    (i) Calculate the magnitude of the gravitational force between the Earth and the Moon. **2**

    (ii) Hence calculate the tangential speed of the Moon in its orbit around the Earth. **2**

    (iii) Define the term *gravitational potential* at a point in space. **1**

    (iv) Calculate the potential energy of the Moon in its orbit. **2**

    (v) Hence calculate the total energy of the Moon in its orbit. **2**

  (*b*) (i) Derive an expression for the escape velocity from the surface of an astronomical body. **2**

    (ii) Calculate the escape velocity from the surface of the Moon. **2**

(13)

Marks

4. (*a*) State what is meant by *simple harmonic motion*. **1**

(*b*) The motion of a piston in a car engine closely approximates to simple harmonic motion.

In a typical engine, the top of a piston moves up and down between points A and B, a distance of 0·10 m, as shown in Figure 4.

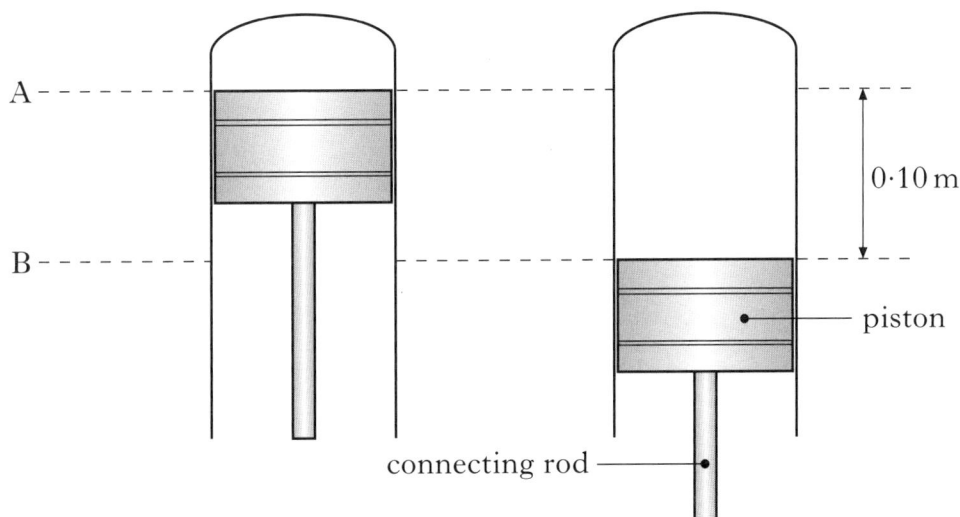

Figure 4

The frequency of the piston's motion is 100 Hz.

Write down an equation which describes how the displacement of the piston from its central position varies with time. Numerical values are required. **2**

(*c*) Calculate the maximum acceleration of the piston. **2**

(*d*) The mass of the piston is 0·48 kg.

Calculate the maximum force applied to the piston by the connecting rod. **2**

(*e*) Calculate the maximum kinetic energy of the piston. **2**

(9)

[Turn over

Marks

5. (a) Figure 5 shows a point charge of +5·1 nC.

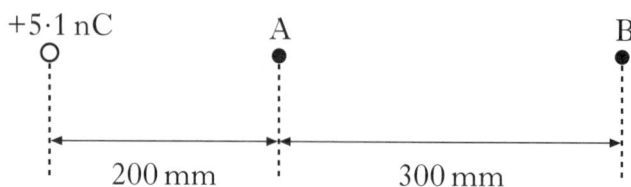

+5·1 nC   A   B
○   ●   ●

|←——— 200 mm ———→|←——— 300 mm ———→|

Figure 5

Point A is a distance of 200 mm from the point charge.
Point B is a distance of 300 mm from point A as shown in Figure 5.

   (i) Show that the potential at point A is 230 V.   1

   (ii) Calculate the potential difference between A and B.   2

(b) A conducting sphere on an insulating support is some distance away from a negatively charged rod as shown in Figure 6.

charged rod

conducting sphere

insulating support

Figure 6

Using diagrams, or otherwise, describe a procedure to charge the sphere positively by induction.   2

Marks

5. (continued)

(c) A charged oil drop of mass $1 \cdot 2 \times 10^{-14}$ kg is stationary between two horizontal parallel plates.

There is a potential difference of $4 \cdot 9$ kV between the parallel plates.

The plates are 80 mm apart as shown in Figure 7.

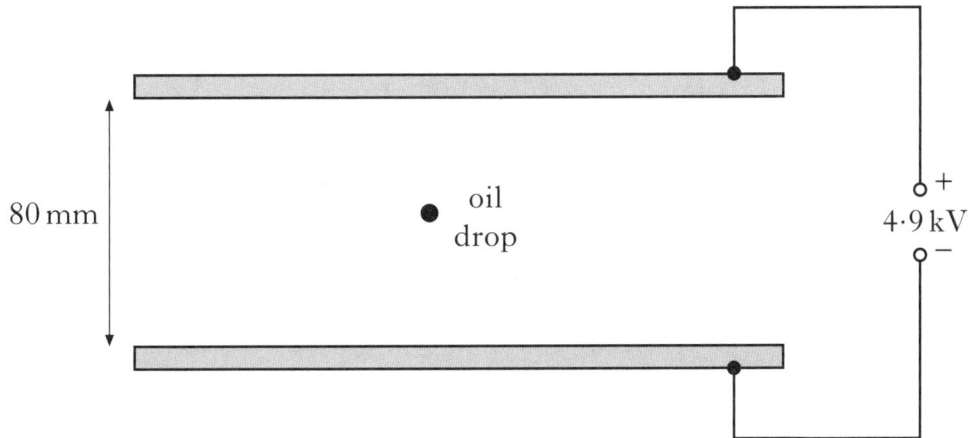

Figure 7

    (i) Draw a labelled diagram to show the forces acting on the oil drop.   **1**

    (ii) Calculate the charge on the oil drop.   **3**

    (iii) How many excess electrons are on the oil drop?   **1**

(d) The results of Millikan's oil drop experiment led to the idea of quantisation of charge.

A down quark has a charge of $-5 \cdot 3 \times 10^{-20}$ C. Explain how this may conflict with Millikan's conclusion.   **1**

    (11)

[Turn over

Marks

6. The shape of the Earth's magnetic field is shown in Figure 8.

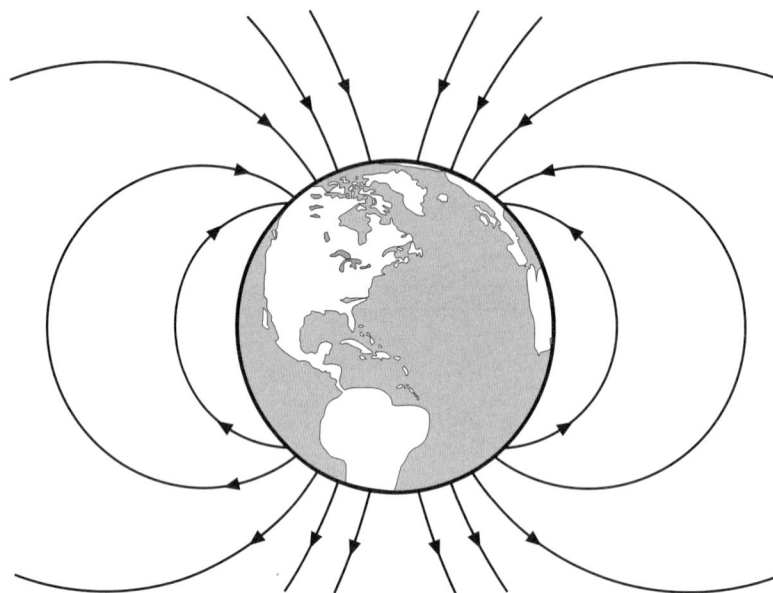

Figure 8

At a particular location in Scotland the field has a magnitude of 5.0×10^{-5} T directed into the Earth's surface at an angle of $69°$ as shown in Figure 9.

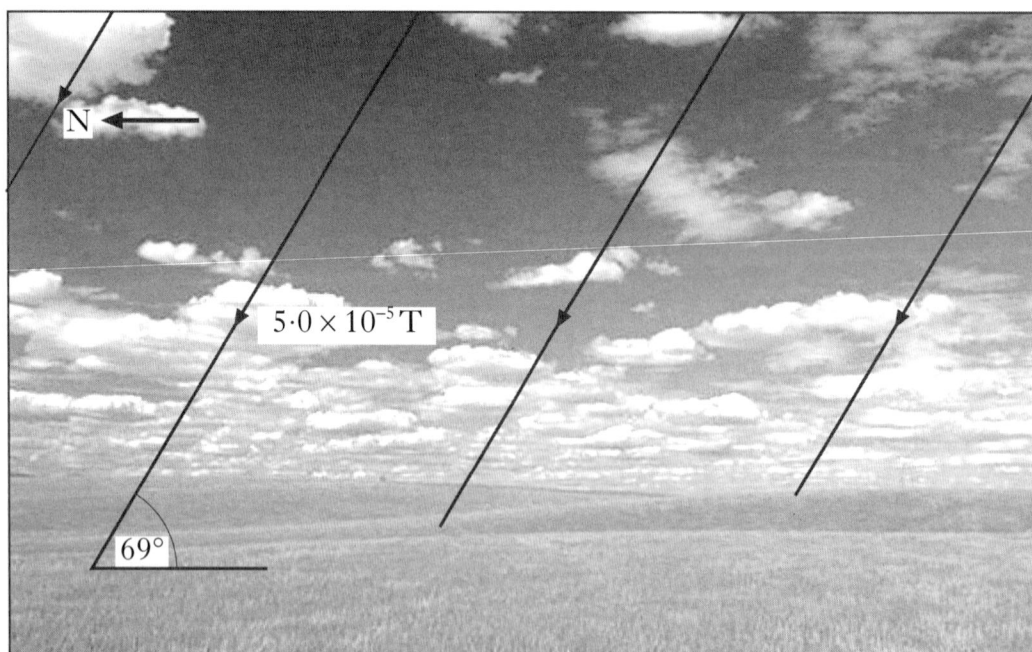

N ←

5.0×10^{-5} T

$69°$

Figure 9

(a) Show that the component of the field perpendicular to the Earth's surface is 4.7×10^{-5} T. 　　1

Marks

6. (continued)

(b) At this location a student sets up a circuit containing a straight length of copper wire lying horizontally in the North – South direction as shown in Figure 10.

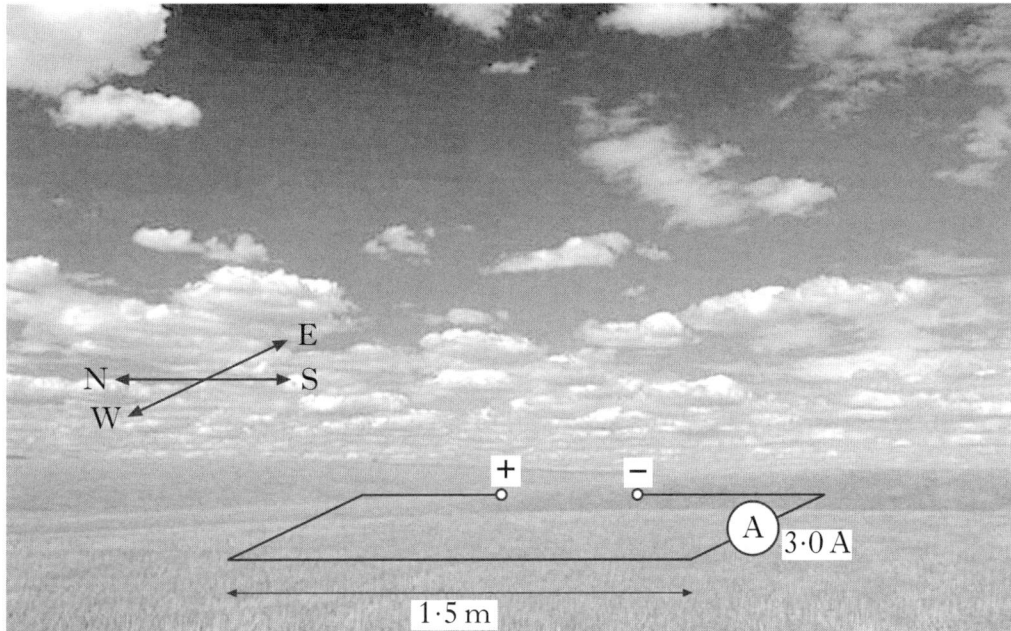

Figure 10

The length of the wire is 1·5 m and the current in the circuit is 3·0 A.

  (i) Calculate the magnitude of the force acting on the wire due to the Earth's magnetic field.   2

  (ii) State the direction of this force.   1

(c) The wire is now tilted through an angle of 69° so that it is parallel to the direction of the Earth's magnetic field.

Determine the force on the wire due to the Earth's magnetic field.   1

(d) A long straight current carrying wire produces a magnetic field. The current in this wire is 3·0 A.

  (i) Calculate the distance from the wire at which the magnitude of the magnetic field is $5·0 \times 10^{-5}$ T.   2

  (ii) Describe the shape of this magnetic field.   1

                                                  (8)

[Turn over

Marks

7. (a) Figure 11 shows a d.c. power supply in series with a switch, lamp and inductor.

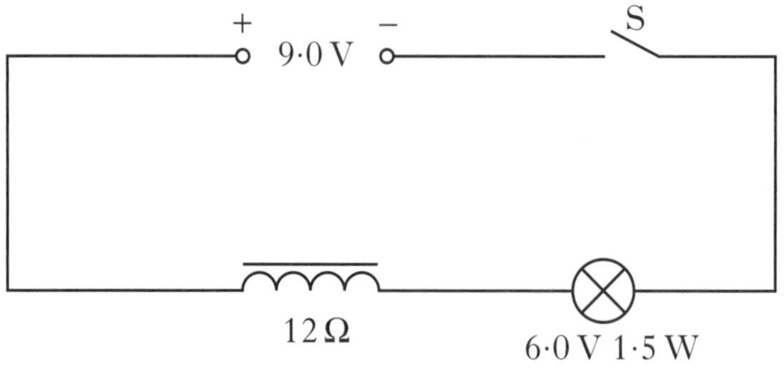

Figure 11

The inductor consists of a coil of wire with a resistance of $12\,\Omega$.
The lamp is rated at $6{\cdot}0\,V\ 1{\cdot}5\,W$.
The $9{\cdot}0\,V$ d.c. power supply has negligible internal resistance.

   (i) Explain why the lamp does not reach its maximum brightness immediately after the switch is closed.   2

  (ii) When the lamp reaches its maximum brightness it is operating at its stated power rating.

      Calculate the current in the circuit.   1

 (iii) The maximum energy stored in the inductor is $75\,mJ$.

      Calculate the inductance of the inductor.   2

 (iv) The inductor in Figure 11 is replaced with another inductor which has the same type of core and wire, but with twice as many turns.

      State the effect this has on:

      (A) the maximum current;

      (B) the time to reach maximum current.   2

Marks

7. (continued)

(b) Figure 12 shows a neon lamp connected to an inductor, switch and a 1·5 V cell.

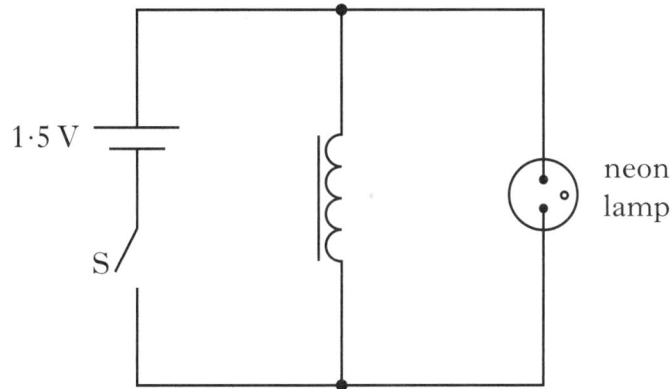

Figure 12

A neon lamp needs a potential difference of at least 80 V across it before it lights.

The switch is closed for 5 seconds.

The switch is then opened and the neon lamp flashes **briefly**.

Explain this observation.   2

(9)

[Turn over

Marks

8. An electron travelling at $9 \cdot 5 \times 10^7 \, \text{m s}^{-1}$ enters a uniform magnetic field B at an angle of $60°$ as shown in Figure 13.

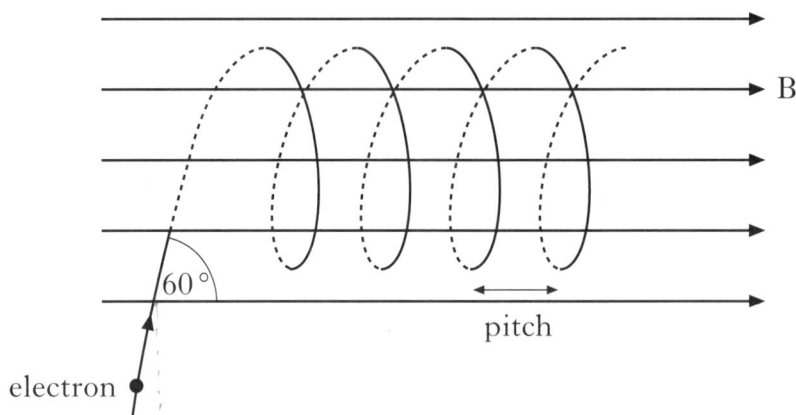

Figure 13

The electron moves in a helical path in the magnetic field.

(a) (i) Calculate the component of the electron's initial velocity:

    (A) parallel to the magnetic field;   **1**

    (B) perpendicular to the magnetic field.   **1**

  (ii) By making reference to **both** components, explain why the electron moves in a helical path.   **2**

(b) (i) The magnetic field has a magnetic induction of $0 \cdot 22 \, \text{T}$.

    Show that the radius of the helix is $2 \cdot 1 \times 10^{-3} \, \text{m}$.   **2**

  (ii) Calculate the time taken for the electron to make one complete revolution.   **2**

  (iii) The distance between adjacent loops in the helix is called the pitch as shown in Figure 13.

     Calculate the pitch of the helix.   **2**

(c) A proton enters the magnetic field with the same initial speed and direction as the electron shown in Figure 13. The magnetic field remains unchanged.

State **two** ways that the path of the proton in the magnetic field is different from the path of the electron.   **2**

           (12)

Marks

9. (a) A water wave travels with a speed of $0.060\,\mathrm{m\,s^{-1}}$ in the positive x direction. Figure 14 represents the water wave at one instant in time.

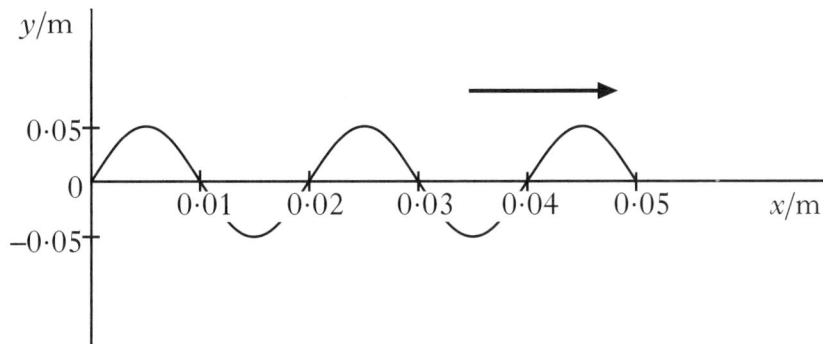

Figure 14

Write down an equation for the vertical displacement y of a point on the water surface in terms of the horizontal displacement x and time t.

Numerical values are required.   2

(b) Write down an equation for an identical wave travelling in the opposite direction.   1

(c) The amplitude of the wave gradually decreases.

Calculate the amplitude of the water wave when the intensity of the wave has decreased by 50%.   2

    (5)

[Turn over

Marks

10. (*a*) A thin coating of magnesium fluoride is applied to the surface of a camera lens.

Figure 15 shows an expanded view of this coating on the glass lens.

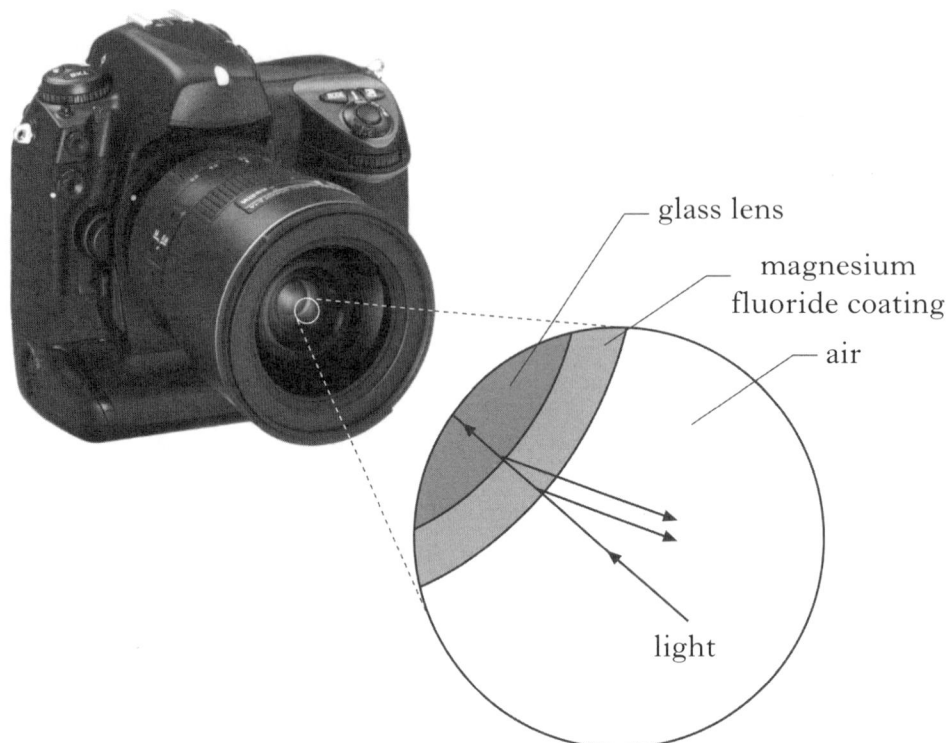

Figure 15

Monochromatic light is incident on the lens and some light reflects from the front and rear surfaces of the coating as shown in Figure 15.

  (i) State the phase change undergone by the light reflected from:

    (A) the front surface of the coating;

    (B) the rear surface of the coating.   **1**

  (ii) Explain, in terms of optical path difference, why this coating can make the lens non-reflecting for a particular wavelength of light.   **2**

  (iii) Why is it desirable that camera lenses should reflect very little light?   **1**

  (iv) A particular lens has a magnesium fluoride coating of thickness $1{\cdot}05 \times 10^{-7}$m.

    Calculate the wavelength of light for which this lens is non-reflecting.   **2**

Marks

10. (continued)

(b) A thin air wedge is formed between two glass plates which are in contact at one end and separated by a thin metal wire at the other end.

Figure 16 shows sodium light being reflected down onto the air wedge. A travelling microscope is used to view the resulting interference pattern.

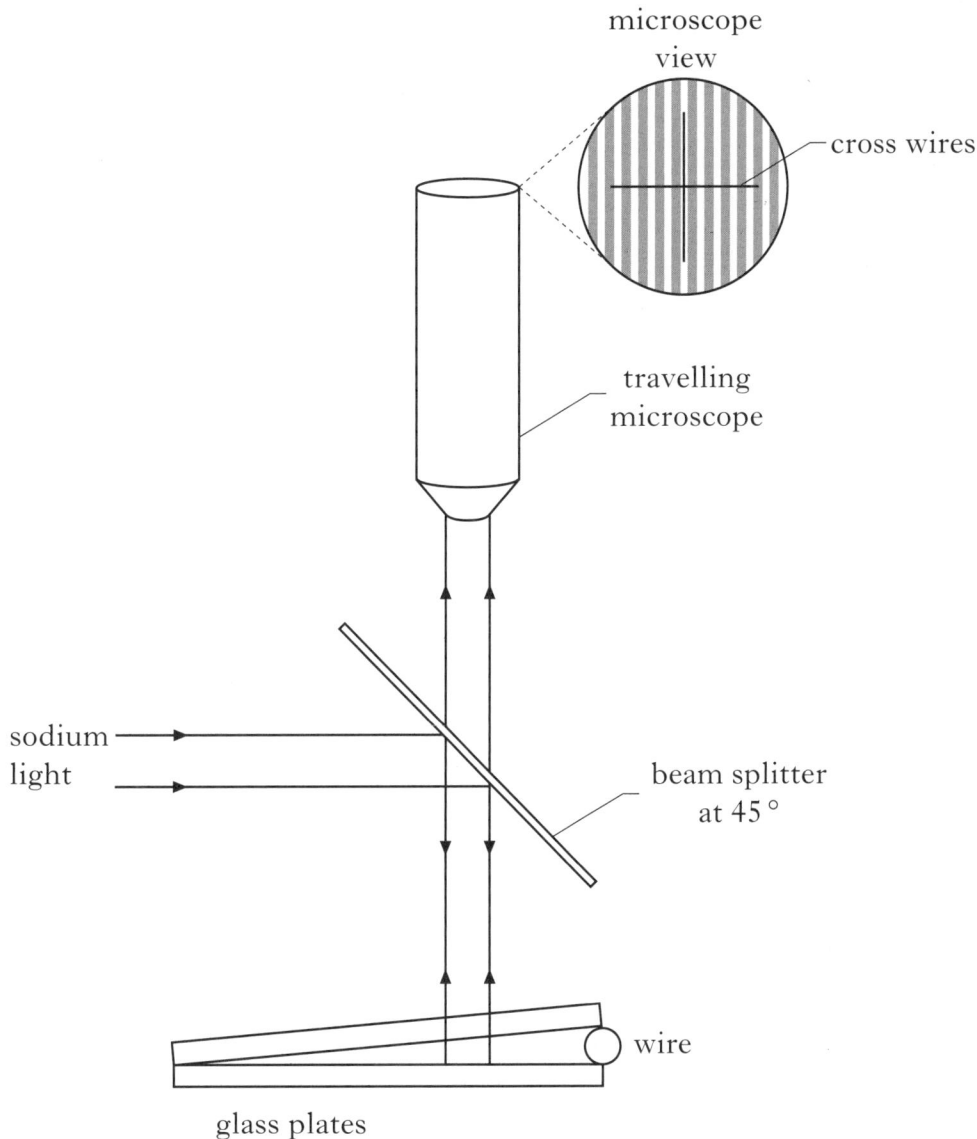

Figure 16

Explain how the diameter of the wire is determined using measurements obtained with this apparatus.

Assume the sodium light is monochromatic.

Your answer should include:

- the measurements required
- any data required
- the equation used.

2

(8)

Marks

11. The apparatus shown in Figure 17 is set up to measure the speed of transverse waves on a stretched string.

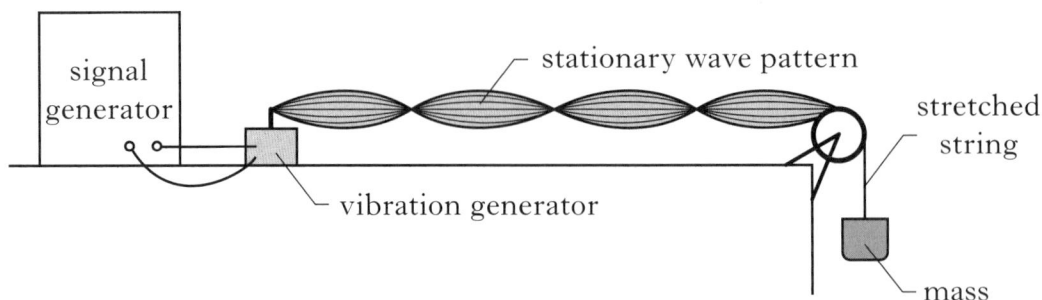

Figure 17

The following data are obtained.

Distance between adjacent nodes = (0.150 ± 0.005) m
Frequency of signal generator = (250 ± 10) Hz

(a) Show that the wave speed is $75\,\text{m s}^{-1}$. 2

(b) Calculate the absolute uncertainty in this value for the wave speed. Express your answer in the form $(75 \pm \quad)\,\text{m s}^{-1}$. 3

(c) (i) In an attempt to reduce the absolute uncertainty, the frequency of the signal generator is increased to (500 ± 10) Hz. Explain why this will **not** result in a reduced absolute uncertainty. 1

(ii) State how the absolute uncertainty in wave speed could be reduced. 1

(7)

[END OF QUESTION PAPER]

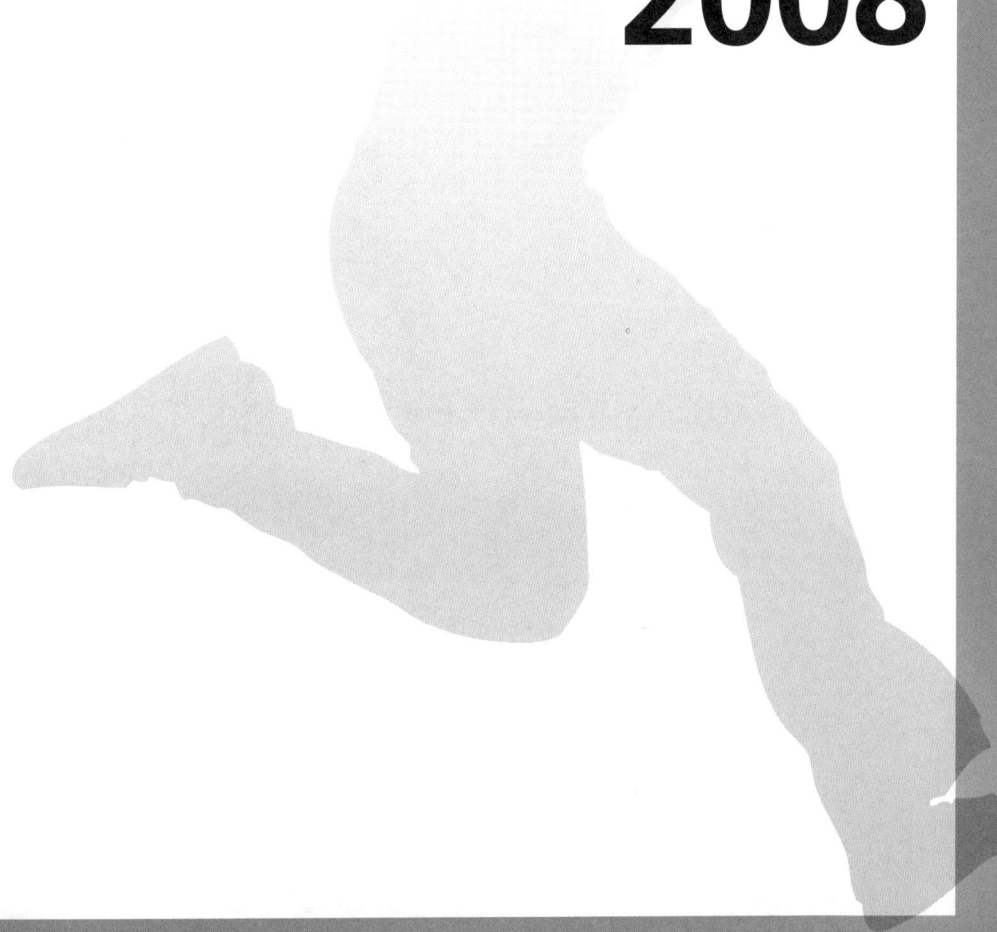

ADVANCED HIGHER

2008

[BLANK PAGE]

X069/701

NATIONAL QUALIFICATIONS 2008	FRIDAY, 23 MAY 1.00 PM – 3.30 PM	PHYSICS ADVANCED HIGHER

Reference may be made to the Physics Data Booklet.

Answer **all** questions.

Any necessary data may be found in the Data Sheet on page three.

Care should be taken to give an appropriate number of significant figures in the final answers to calculations.

Square-ruled paper (if used) should be placed inside the front cover of the answer book for return to the Scottish Qualifications Authority.

XXSQA

[BLANK PAGE]

DATA SHEET
COMMON PHYSICAL QUANTITIES

Quantity	Symbol	Value	Quantity	Symbol	Value
Gravitational acceleration on Earth	g	9.8 m s^{-2}	Mass of electron	m_e	$9.11 \times 10^{-31} \text{ kg}$
Radius of Earth	R_E	$6.4 \times 10^6 \text{ m}$	Charge on electron	e	$-1.60 \times 10^{-19} \text{ C}$
Mass of Earth	M_E	$6.0 \times 10^{24} \text{ kg}$	Mass of neutron	m_n	$1.675 \times 10^{-27} \text{ kg}$
Mass of Moon	M_M	$7.3 \times 10^{22} \text{ kg}$	Mass of proton	m_p	$1.673 \times 10^{-27} \text{ kg}$
Radius of Moon	R_M	$1.7 \times 10^6 \text{ m}$	Mass of alpha particle	m_α	$6.645 \times 10^{-27} \text{ kg}$
Mean Radius of Moon Orbit		$3.84 \times 10^8 \text{ m}$	Charge on alpha particle		$3.20 \times 10^{-19} \text{ C}$
Universal constant of gravitation	G	$6.67 \times 10^{-11} \text{ m}^3 \text{ kg}^{-1} \text{ s}^{-2}$	Planck's constant	h	$6.63 \times 10^{-34} \text{ J s}$
Speed of light in vacuum	c	$3.0 \times 10^8 \text{ m s}^{-1}$	Permittivity of free space	ε_0	$8.85 \times 10^{-12} \text{ F m}^{-1}$
Speed of sound in air	v	$3.4 \times 10^2 \text{ m s}^{-1}$	Permeability of free space	μ_0	$4\pi \times 10^{-7} \text{ H m}^{-1}$

REFRACTIVE INDICES
The refractive indices refer to sodium light of wavelength 589 nm and to substances at a temperature of 273 K.

Substance	Refractive index	Substance	Refractive index
Diamond	2·42	Glycerol	1·47
Glass	1·51	Water	1·33
Ice	1·31	Air	1·00
Perspex	1·49	Magnesium Fluoride	1·38

SPECTRAL LINES

Element	Wavelength/nm	Colour	Element	Wavelength/nm	Colour
Hydrogen	656	Red	Cadmium	644	Red
	486	Blue-green		509	Green
	434	Blue-violet		480	Blue
	410	Violet			
	397	Ultraviolet		*Lasers*	
	389	Ultraviolet	Element	Wavelength/nm	Colour
Sodium	589	Yellow	Carbon dioxide	9550 } 10590 }	Infrared
			Helium-neon	633	Red

PROPERTIES OF SELECTED MATERIALS

Substance	Density/ kg m^{-3}	Melting Point/ K	Boiling Point/ K	Specific Heat Capacity/ J kg^{-1} K^{-1}	Specific Latent Heat of Fusion/ J kg^{-1}	Specific Latent Heat of Vaporisation/ J kg^{-1}
Aluminium	2.70×10^3	933	2623	9.02×10^2	3.95×10^5	....
Copper	8.96×10^3	1357	2853	3.86×10^2	2.05×10^5	....
Glass	2.60×10^3	1400	....	6.70×10^2	....	....
Ice	9.20×10^2	273	....	2.10×10^3	3.34×10^5	....
Glycerol	1.26×10^3	291	563	2.43×10^3	1.81×10^5	8.30×10^5
Methanol	7.91×10^2	175	338	2.52×10^3	9.9×10^4	1.12×10^6
Sea Water	1.02×10^3	264	377	3.93×10^3	....	....
Water	1.00×10^3	273	373	4.19×10^3	3.34×10^5	2.26×10^6
Air	1·29	....	....	....	....	....
Hydrogen	9.0×10^{-2}	14	20	1.43×10^4	....	4.50×10^5
Nitrogen	1·25	63	77	1.04×10^3	....	2.00×10^5
Oxygen	1·43	55	90	9.18×10^2	....	2.40×10^5

The gas densities refer to a temperature of 273 K and a pressure of 1.01×10^5 Pa.

Marks

1. A centrifuge is used to separate out small particles suspended in a liquid. Figure 1 shows the rotating part of the centrifuge which includes two test tubes containing the liquid.

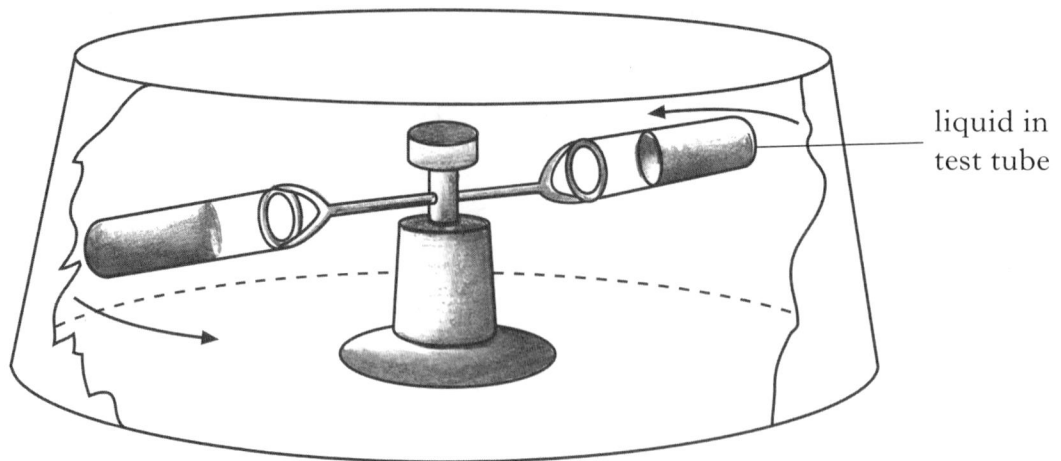

liquid in test tube

Figure 1

The rotating part starts from rest and reaches a maximum angular velocity of $1200\,\text{rad s}^{-1}$ in a time of 4 seconds.

The average moment of inertia of the rotating part is $5\cdot1 \times 10^{-4}\,\text{kg m}^2$.

(a)   (i)   Calculate the angular acceleration of the rotating part.   **2**

     (ii)   Calculate the average unbalanced torque applied during this time.   **2**

     (iii)   How many **revolutions** are made during this time?   **3**

(b)   Figure 2 shows an overhead view of the rotating part.

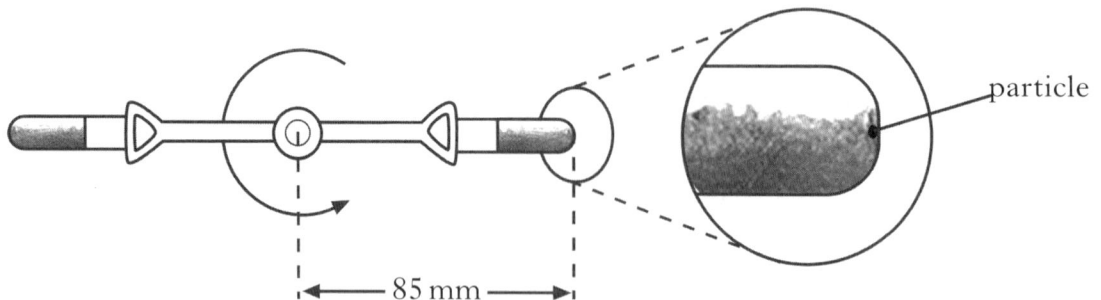

particle

85 mm

Figure 2

The expanded view shows the position of a single particle of mass $5\cdot3 \times 10^{-6}\,\text{kg}$.

   (i)   Calculate the central force acting on the particle at the maximum angular velocity.   **2**

   (ii)   What provides the central force acting on this particle?   **1**

Marks

1. (continued)

(c) At rest the test tubes in the centrifuge are in a vertical position as shown in Figure 3.

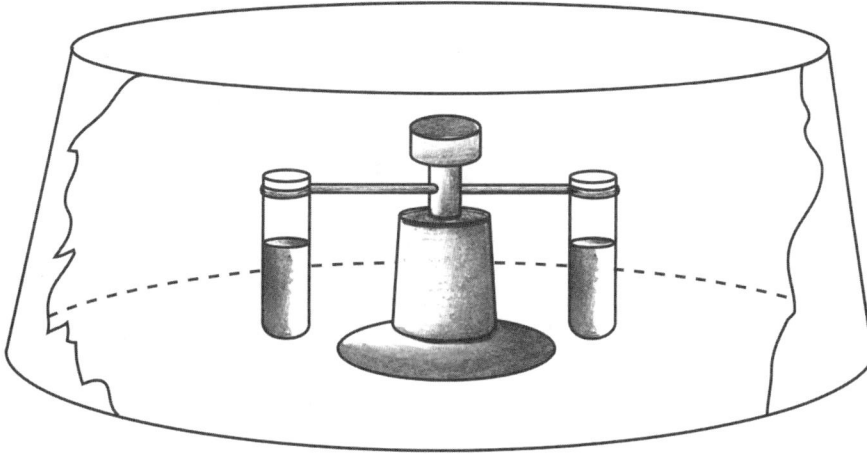

Figure 3

Does the moment of inertia of the rotating part increase, decrease, or stay the same during the acceleration of the rotating part? Justify your answer.　　2

(12)

[Turn over

Marks

2. (a) The gravitational field strength g on the surface of Mars is $3\cdot7\,\text{N}\,\text{kg}^{-1}$.
The mass of Mars is $6\cdot4 \times 10^{23}\,\text{kg}$.
Show that the radius of Mars is $3\cdot4 \times 10^{6}\,\text{m}$.   2

(b) (i) A satellite of mass m has an orbit of radius R. Show that the angular velocity ω of the satellite is given by the expression

$$\omega = \sqrt{\frac{GM}{R^{3}}}$$

where the symbols have their usual meanings.   2

(ii) A satellite remains above the same point on the equator of Mars as the planet spins on its axis.

Figure 4 shows this satellite orbiting at a height of $1\cdot7 \times 10^{7}\,\text{m}$ above the Martian surface.

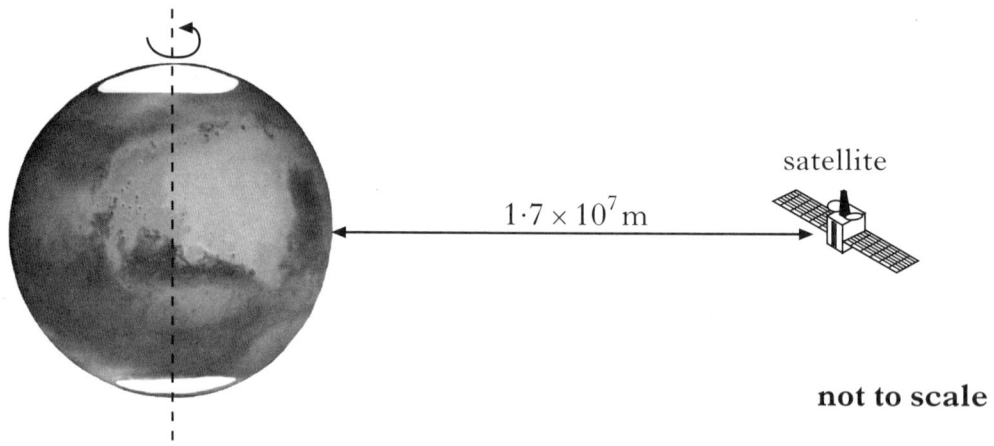

not to scale

Figure 4

Calculate the angular velocity of the satellite.   2

(iii) Calculate the length of one Martian day.   2

(c) The following table gives data about three planets orbiting the Sun.

Planet	Radius R of orbit around the Sun/10^{9}m	Orbit period T around the Sun/years
Venus	108	0·62
Mars	227	1·88
Jupiter	780	12·0

Use **all** the data to show that T^{2} is directly proportional to R^{3} for these three planets.   3

    (11)

Marks

3. A simple pendulum consists of a lead ball on the end of a long string as shown in Figure 5.

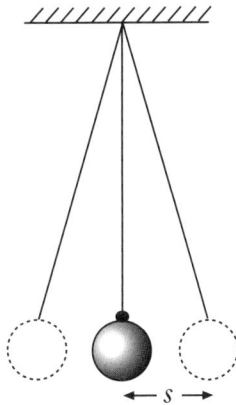

Figure 5

The ball moves with simple harmonic motion. At time t the displacement s of the ball is given by the expression

$$s = 2 \cdot 0 \times 10^{-2} \cos 4 \cdot 3t$$

where s is in metres and t in seconds.

(a) (i) State the definition of *simple harmonic motion*.   1

    (ii) Calculate the period of the pendulum.   2

(b) Calculate the maximum speed of the ball.   2

(c) The mass of the ball is $5 \cdot 0 \times 10^{-2}$ kg and the string has negligible mass.

   Calculate the total energy of the pendulum.   2

(d) The period T of a pendulum is given by the expression

$$T = 2\pi \sqrt{\frac{L}{g}}$$

   where L is the length of the pendulum.

   Calculate the length of this pendulum.   2

(e) In the above case, the assumption has been made that the motion is not subject to *damping*.

   State what is meant by *damping*.   1

                                                            (10)

[Turn over

Marks

4. (a) Electrons can exhibit wave-like behaviour. Give **one** example of evidence which supports this statement.   1

   (b) The Bohr model of the hydrogen atom suggests a nucleus with an electron occupying one of a series of stable orbits.

   A nucleus and the first two stable orbits are shown in Figure 6.

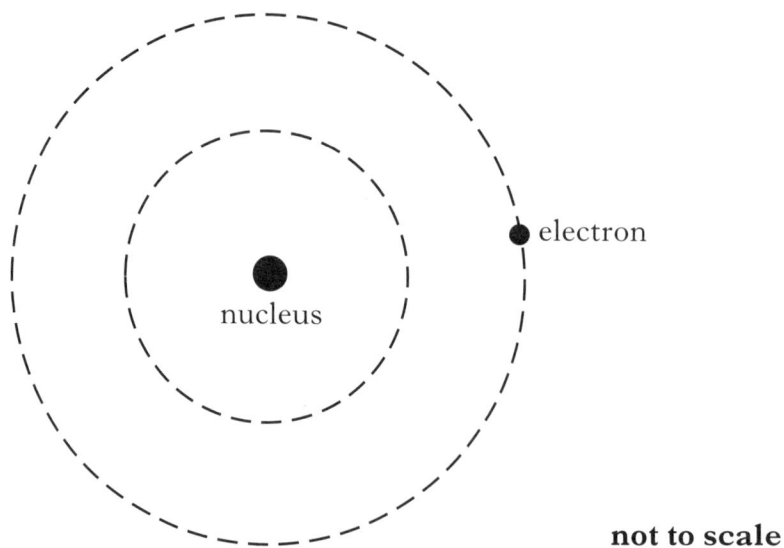

not to scale

Figure 6

   (i) Calculate the angular momentum of the electron in the second stable orbit.   2

   (ii) Starting with the relationship

$$mrv = \frac{nh}{2\pi}$$

   show that the circumference of the second stable orbit is equal to two electron wavelengths.   2

   (iii) The circumference of the second stable orbit is $1 \cdot 3 \times 10^{-9}$ m.

   Calculate the speed of the electron in this orbit.   2

                                                                                                                   (7)

Marks

5. (*a*) Two point charges Q_1 and Q_2 each has a charge of $-4\cdot0\,\mu C$. The charges are $0\cdot60\,m$ apart as shown in Figure 7.

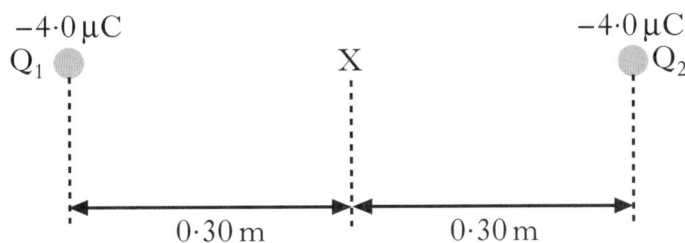

$-4\cdot0\,\mu C$ Q_1 X $-4\cdot0\,\mu C$ Q_2

$0\cdot30\,m$ $0\cdot30\,m$

Figure 7

    (i) Draw a diagram to show the electric field lines between charges Q_1 and Q_2.   **1**

    (ii) Calculate the electrostatic potential at point X, midway between the charges.   **2**

(*b*) A third point charge Q_3 is placed near the two charges as shown in Figure 8.

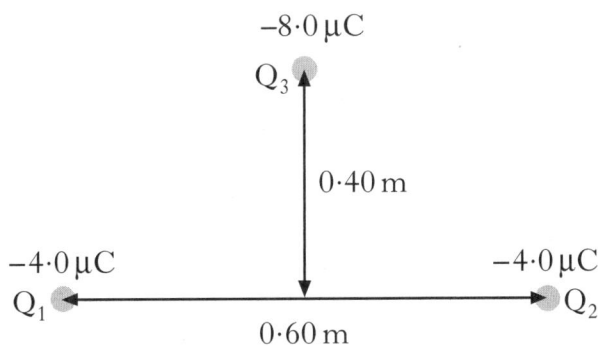

$-8\cdot0\,\mu C$

Q_3

$0\cdot40\,m$

$-4\cdot0\,\mu C$ Q_1 $-4\cdot0\,\mu C$ Q_2

$0\cdot60\,m$

Figure 8

    (i) Show that the force between charges Q_1 and Q_3 is $1\cdot2\,N$.   **2**

    (ii) Calculate the **magnitude** and **direction** of the resultant force on charge Q_3 due to charges Q_1 and Q_2.   **2**

           (7)

[Turn over

Marks

6. A student investigates the relationship between the force exerted on a wire in a magnetic field and the current in the wire.

A pair of magnets is fixed to a yoke and placed on a top pan Newton balance. A rigid copper wire is suspended between the poles of the magnets. The wire is fixed at 90° to the magnetic field, as shown in Figure 9.

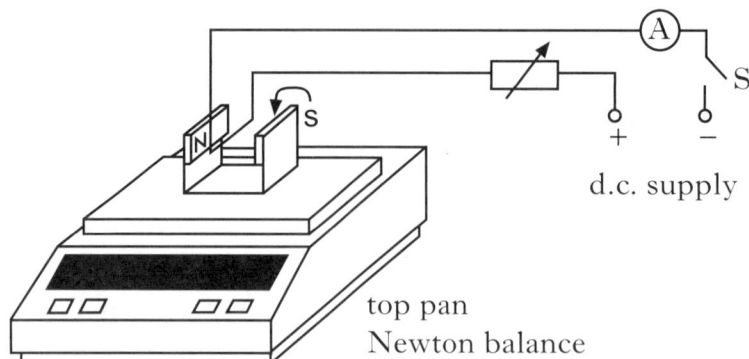

Figure 9

With switch S open the balance is set to zero.

Switch S is closed. The resistor is adjusted and the force recorded for several values of current.

The results are given in the table below.

Current/A	0·50	1·00	1·50	2·00	2·50
Force/10^{-3} N	0·64	0·85	2·56	3·07	3·87

The uncertainty in the current is $\pm\, 0\cdot01$ A.
The uncertainty in the force is $\pm\, 0\cdot03 \times 10^{-3}$ N.

Figure 10, on *Page eleven*, shows the corresponding graph with the best fit straight line for the results.

(a)   (i)   Show that the gradient of the line is $1\cdot7 \times 10^{-3}\,\text{N A}^{-1}$.   **1**

     (ii)   Calculate the absolute uncertainty in the gradient of the line.   **3**

    (iii)   The length of wire in the magnetic field is 52 mm. Use the information obtained from the graph to calculate the magnitude of the magnetic induction.

         The uncertainty in the magnetic induction is **not** required.   **2**

(b)   In the student's evaluation it is stated that the line does not pass through the origin.

    (i)   Suggest a possible reason for this.   **1**

    (ii)   Suggest **one** improvement to the experiment to reduce the absolute uncertainty in the gradient of the line.   **1**

                                                   (8)

6. (continued)

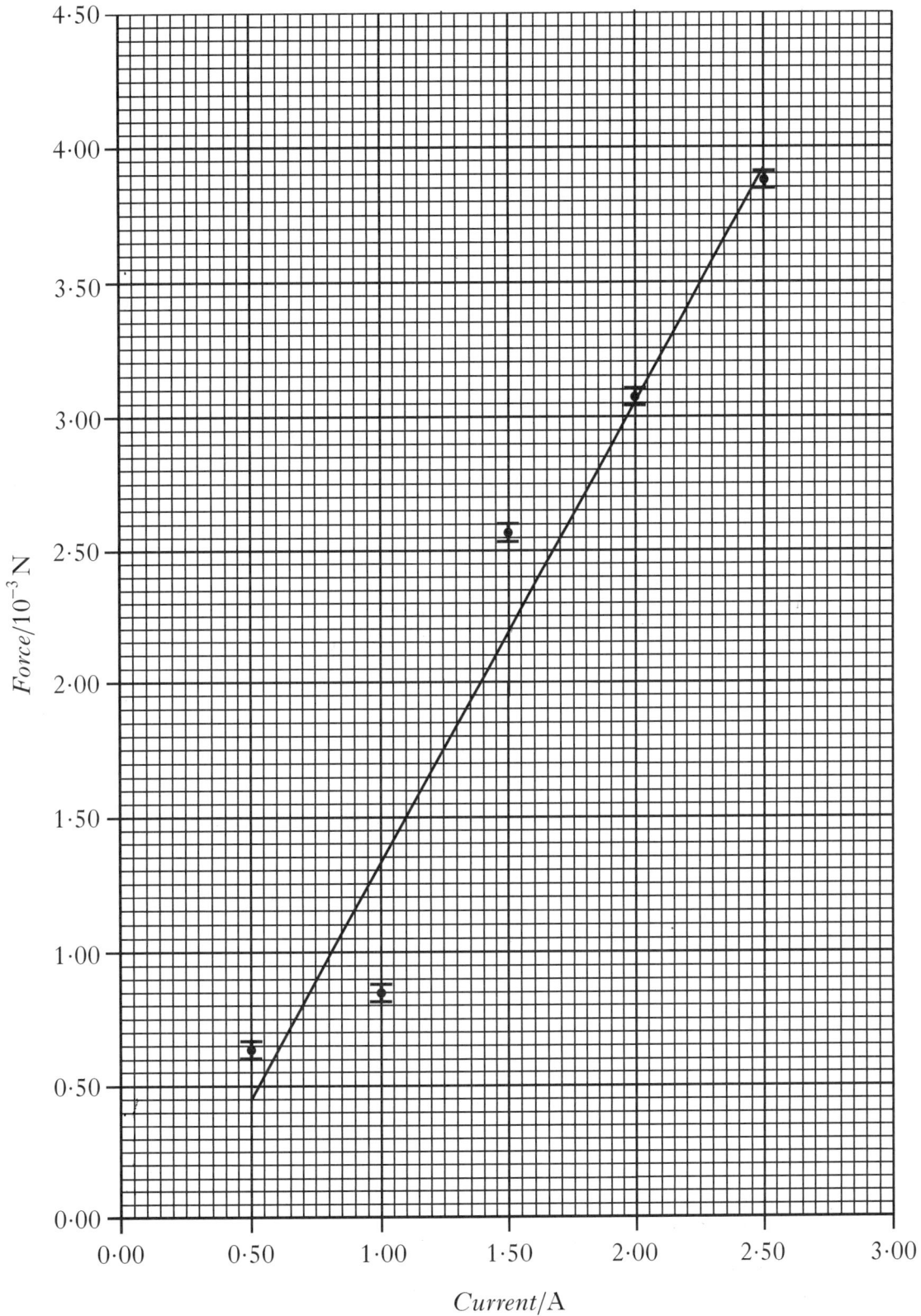

Figure 10

[Turn over

Marks

7. An inductor of negligible resistance is connected in the circuit shown in Figure 11.

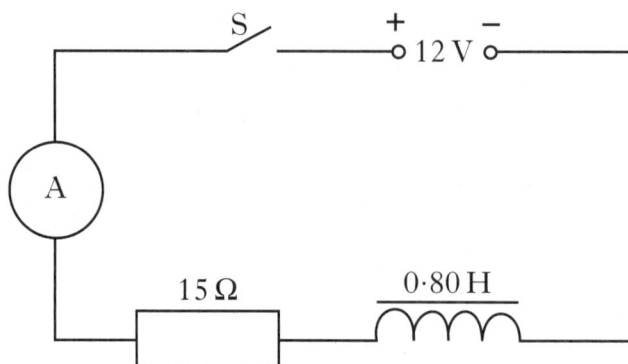

Figure 11

(a) The inductor has an inductance of 0·80 H.

Switch S is closed.

   (i) Explain why there is a time delay before the current reaches its maximum value.   **1**

   (ii) Calculate the maximum current in the circuit.   **2**

   (iii) Calculate the maximum energy stored in the inductor.   **2**

   (iv) Calculate the rate of change of current when the current in the circuit is 0·12 A.   **3**

(b) Switch S is opened and the iron core is removed from the inductor. Switch S is now closed.

   (i) Will the maximum current be bigger, smaller or the same as the maximum current calculated in (a)(ii)?   **1**

   (ii) Explain any change in the time delay to reach the maximum current.   **2**

   (iii) Explain why the maximum energy stored in the inductor is less than in (a)(iii).   **1**

(c) The iron core is replaced in the inductor. The d.c. supply is replaced with a variable frequency supply as shown in Figure 12.

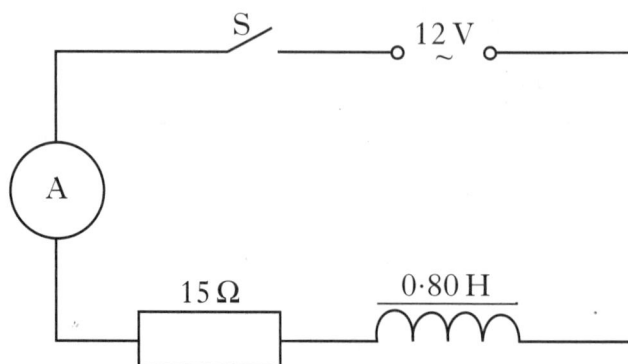

Figure 12

Sketch a graph to show how the current in the circuit varies with the frequency of the supply. Numerical values are not required.   **1**

(13)

Marks

8. (a) Two protons are separated by a distance of $22\,\mu\text{m}$.

   (i) Show by calculation that the gravitational force between these protons is negligible compared to the electrostatic force. 4

   (ii) Why is the strong force negligible between these protons? 1

   (b) A particle of charge q travels directly towards a fixed stationary particle of charge Q.

   At a large distance from charge Q the moving particle has an initial velocity v.

   The moving particle momentarily comes to rest at a distance of closest approach r_c as shown in Figure 13.

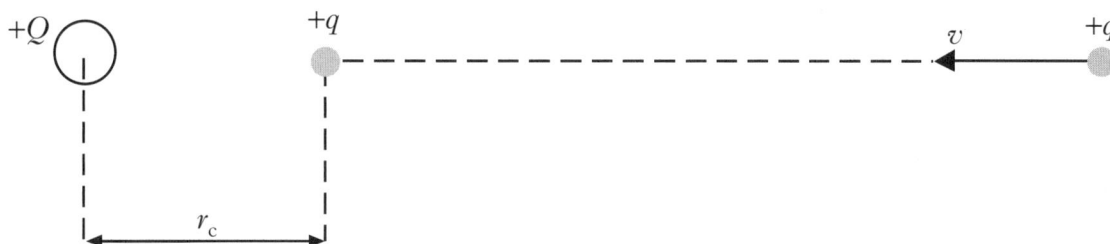

Figure 13

Show that the initial velocity of the moving particle is given by

$$v = \sqrt{\frac{qQ}{2\pi\varepsilon_0 m r_c}}$$

where the symbols have their usual meaning. 2

   (c) An alpha particle is fired towards a target nucleus which is fixed and stationary. The initial velocity of the alpha particle is $9\cdot63 \times 10^6\,\text{m}\,\text{s}^{-1}$ and the distance of closest approach is $1\cdot12 \times 10^{-13}\,\text{m}$.

   (i) Calculate the charge on the target nucleus. 3

   (ii) Calculate the number of protons in the target nucleus. 2

   (iii) The target is the nucleus of an element. Identify this element. 1

   (13)

[Turn over

Marks

9. (*a*) The driver of a sports car approaches a building where an alarm is sounding as shown in Figure 14.

Figure 14

The speed of the car is $25.0 \, \text{m s}^{-1}$ and the frequency of the sound emitted by the alarm is 1250 Hz.

   (i) Explain in terms of wavefronts why the sound heard by the driver does not have a frequency of 1250 Hz. You may wish to include a diagram to support your answer.   2

   (ii) Calculate the frequency of the sound from the alarm heard by the driver.   2

Marks

9. (continued)

(b) The spectrum of light from most stars contains lines corresponding to helium gas.

Figure 15(a) shows the helium spectrum from the Sun.

Figure 15(b) shows the helium spectrum from a distant star.

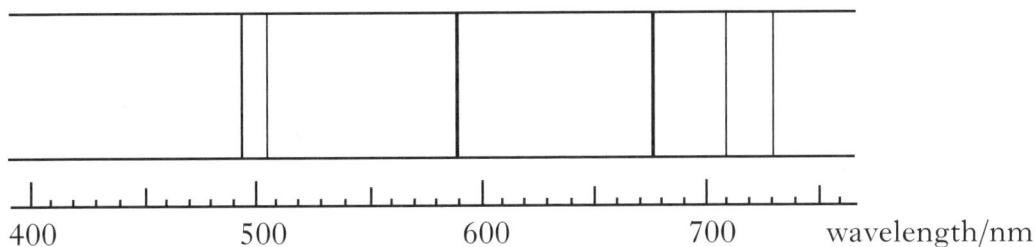

Figure 15(a)

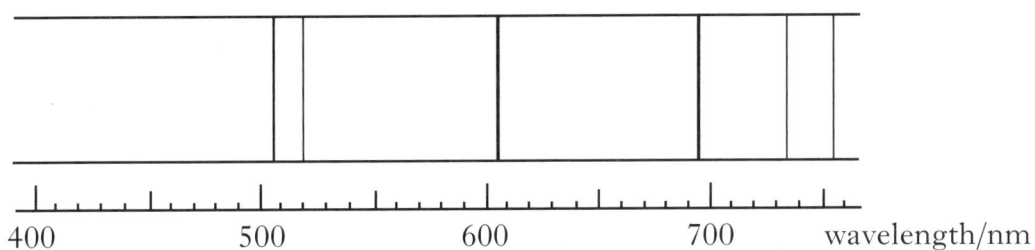

Figure 15(b)

By comparing these spectra, what conclusion can be made about the distant star? Justify your answer.

2

(6)

[Turn over

Marks

10. (*a*) (i) State what is meant by the term *plane polarised light*.   **1**

(ii) Figure 16 shows the refraction of red light at a water-air interface.

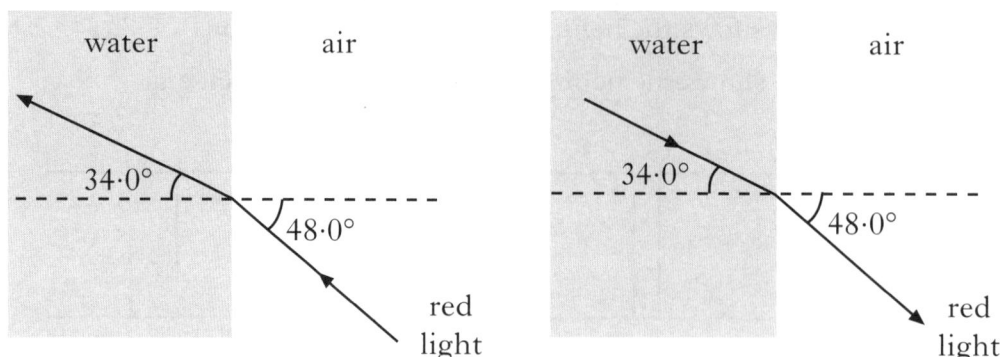

Figure 16

The refractive index *n* for red light travelling from air to water is 1·33. Show that the refractive index μ for red light travelling from **water** to **air** is 0·752.   **1**

(iii) Figure 17 shows a ray of unpolarised red light incident on a water-air interface.

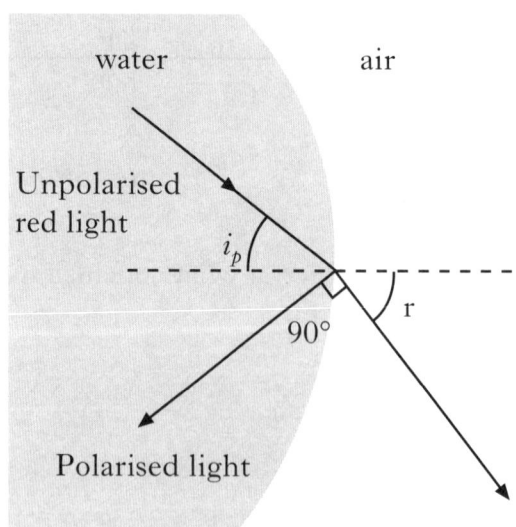

Figure 17

For light travelling from water to air,

$$\mu = \tan i_p$$

where i_p is the Brewster angle.

Calculate the Brewster angle for red light at this water-air interface.   **1**

Marks

10. (continued)

(b) A rainbow is produced when light follows the path in a raindrop as shown in Figure 18.

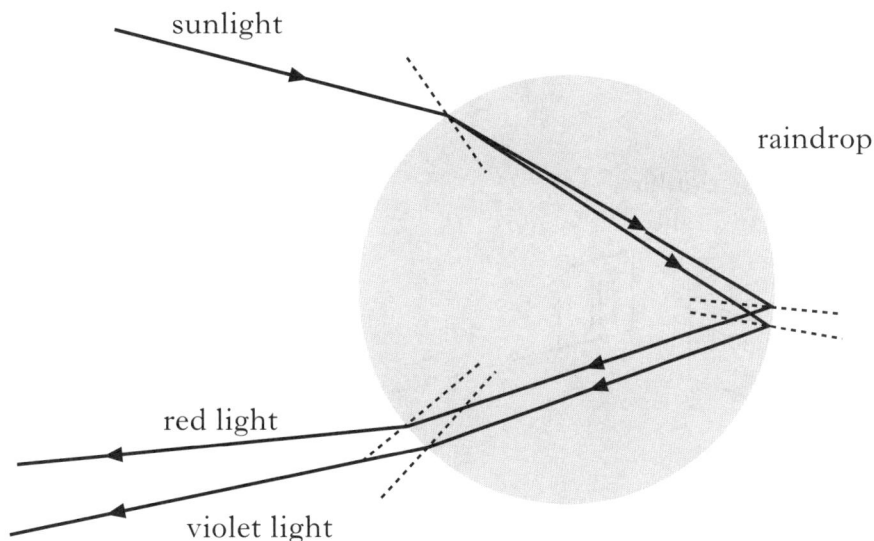

Figure 18

The light emerging from the raindrop is polarised.

The refractive index, μ, at a water to air interface is 0·752 for red light and 0·745 for violet light.

Calculate the difference in Brewster's angle for these two colours.   2

(c) Rainbows produce light that is 96% polarised. A photographer plans to take a photograph of a rainbow. Her camera has a polarising filter in front of the lens as shown in Figure 19.

Figure 19

She directs her camera at the rainbow and slowly rotates the filter to see which is the best image to take.

Describe what happens to the image of the rainbow as she slowly rotates her filter through 180°.   2

(7)

Marks

11. Light from a helium-neon laser is incident on a double slit. A pattern of light and dark fringes is observed on a screen 3·50 m beyond the slits as shown in Figure 20.

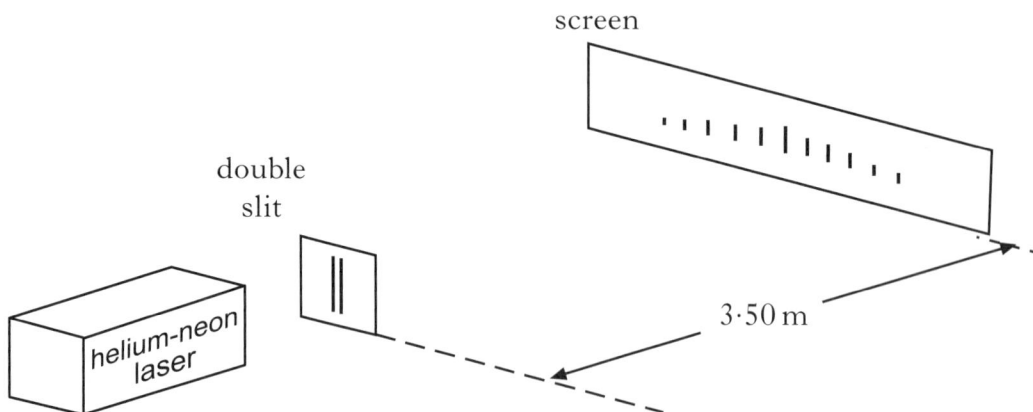

Figure 20

(a) State whether these fringes are caused by division of amplitude or division of wavefront.

1

(b) The distance between two adjacent bright fringes on the screen is 7·20 mm. Calculate the separation of the two slits.

2

(c) The distance between the double slit and screen is increased to 5·50 m. The distance between the fringes is remeasured and the calculation of the slit separation is repeated.

　(i) Explain **one** advantage of moving the screen further away from the double slit.

2

　(ii) State **one** disadvantage of moving the screen further away from the double slit.

1

(6)

[END OF QUESTION PAPER]

ADVANCED HIGHER

2009

[BLANK PAGE]

X069/701

NATIONAL QUALIFICATIONS 2009	TUESDAY, 26 MAY 1.00 PM – 3.30 PM	PHYSICS ADVANCED HIGHER

Reference may be made to the Physics Data Booklet.

Answer **all** questions.

Any necessary data may be found in the Data Sheet on page two.

Care should be taken to give an appropriate number of significant figures in the final answers to calculations.

Square-ruled paper (if used) should be placed inside the front cover of the answer book for return to the Scottish Qualifications Authority.

XSQA

DATA SHEET
COMMON PHYSICAL QUANTITIES

Quantity	Symbol	Value	Quantity	Symbol	Value
Gravitational acceleration on Earth	g	$9.8\ \text{m s}^{-2}$	Mass of electron	m_e	$9.11 \times 10^{-31}\ \text{kg}$
Radius of Earth	R_E	$6.4 \times 10^6\ \text{m}$	Charge on electron	e	$-1.60 \times 10^{-19}\ \text{C}$
Mass of Earth	M_E	$6.0 \times 10^{24}\ \text{kg}$	Mass of neutron	m_n	$1.675 \times 10^{-27}\ \text{kg}$
Mass of Moon	M_M	$7.3 \times 10^{22}\ \text{kg}$	Mass of proton	m_p	$1.673 \times 10^{-27}\ \text{kg}$
Radius of Moon	R_M	$1.7 \times 10^6\ \text{m}$	Mass of alpha particle	m_α	$6.645 \times 10^{-27}\ \text{kg}$
Mean Radius of Moon Orbit		$3.84 \times 10^8\ \text{m}$	Charge on alpha particle		$3.20 \times 10^{-19}\ \text{C}$
Universal constant of gravitation	G	$6.67 \times 10^{-11}\ \text{m}^3\ \text{kg}^{-1}\ \text{s}^{-2}$	Planck's constant	h	$6.63 \times 10^{-34}\ \text{J s}$
Speed of light in vacuum	c	$3.0 \times 10^8\ \text{m s}^{-1}$	Permittivity of free space	ε_0	$8.85 \times 10^{-12}\ \text{F m}^{-1}$
Speed of sound in air	v	$3.4 \times 10^2\ \text{m s}^{-1}$	Permeability of free space	μ_0	$4\pi \times 10^{-7}\ \text{H m}^{-1}$

REFRACTIVE INDICES

The refractive indices refer to sodium light of wavelength 589 nm and to substances at a temperature of 273 K.

Substance	Refractive index	Substance	Refractive index
Diamond	2·42	Glycerol	1·47
Glass	1·51	Water	1·33
Ice	1·31	Air	1·00
Perspex	1·49	Magnesium Fluoride	1·38

SPECTRAL LINES

Element	Wavelength/nm	Colour	Element	Wavelength/nm	Colour
Hydrogen	656	Red	Cadmium	644	Red
	486	Blue-green		509	Green
	434	Blue-violet		480	Blue
	410	Violet	Lasers		
	397	Ultraviolet	Element	Wavelength/nm	Colour
	389	Ultraviolet	Carbon dioxide	9550 } 10590 }	Infrared
Sodium	589	Yellow	Helium-neon	633	Red

PROPERTIES OF SELECTED MATERIALS

Substance	Density/ kg m^{-3}	Melting Point/ K	Boiling Point/ K	Specific Heat Capacity/ $\text{J kg}^{-1}\ \text{K}^{-1}$	Specific Latent Heat of Fusion/ J kg^{-1}	Specific Latent Heat of Vaporisation/ J kg^{-1}
Aluminium	2.70×10^3	933	2623	9.02×10^2	3.95×10^5	. . . .
Copper	8.96×10^3	1357	2853	3.86×10^2	2.05×10^5	. . . .
Glass	2.60×10^3	1400	. . . .	6.70×10^2	. . . .	. . . .
Ice	9.20×10^2	273	. . . .	2.10×10^3	3.34×10^5	. . . .
Glycerol	1.26×10^3	291	563	2.43×10^3	1.81×10^5	8.30×10^5
Methanol	7.91×10^2	175	338	2.52×10^3	9.9×10^4	1.12×10^6
Sea Water	1.02×10^3	264	377	3.93×10^3	. . . .	. . . .
Water	1.00×10^3	273	373	4.19×10^3	3.34×10^5	2.26×10^6
Air	1·29	. . . .	. . . .	. . . .		
Hydrogen	9.0×10^{-2}	14	20	1.43×10^4		4.50×10^5
Nitrogen	1·25	63	77	1.04×10^3		2.00×10^5
Oxygen	1·43	55	90	9.18×10^2		2.40×10^5

The gas densities refer to a temperature of 273 K and a pressure of 1.01×10^5 Pa.

Marks

1. Figure 1A shows a space shuttle shortly after take-off.

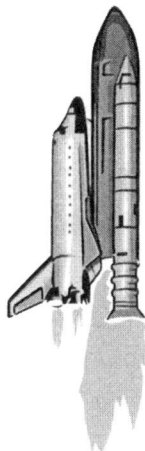

Figure 1A

(a) Immediately after take off, the vertical displacement of the shuttle for part of its journey can be described using the equation

$$s = 3.1t^2 + 4.1t.$$

   (i) Find, by differentiation, the equation for the vertical velocity of the shuttle.   1

   (ii) At what time will the vertical velocity be $72 \, \text{m s}^{-1}$?   2

   (iii) Calculate the vertical linear acceleration during this time.   1

(b) A theory suggests that a burned out star can collapse to form a black hole.

   (i) Explain why a *black hole* appears black.   1

   (ii) Explain the term *escape velocity*.   1

   (iii) Derive an expression for the escape velocity at the surface of a star of mass M and radius r.   2

   (iv) A star collapses to form a black hole of mass $4.58 \times 10^{30} \, \text{kg}$. Even in these extreme conditions the expression in part (iii) applies. Calculate the maximum radius of the black hole.   2

   (v) Calculate the minimum density ρ of this black hole.   3

(13)

[Turn over

Marks

2. An anemometer is an instrument to measure windspeed and is shown in Figure 2A.

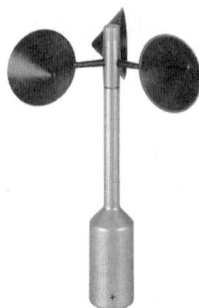

Figure 2A

The anemometer is tested in a wind tunnel.

The calibration graph of the angular velocity, in **revolutions per minute**, against windspeed, in m s^{-1}, is shown in Figure 2B.

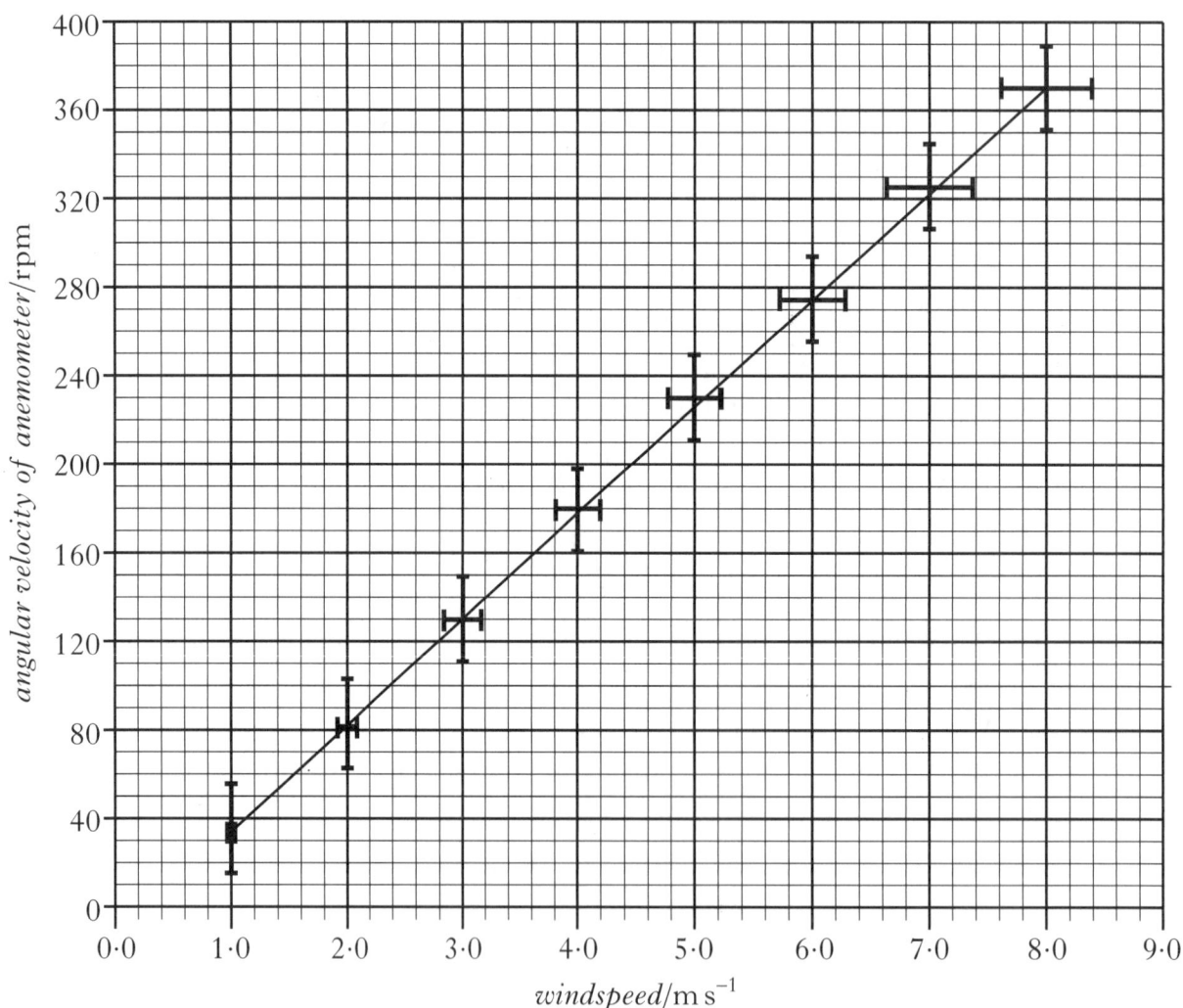

Figure 2B

The calibration graph is found not to go through the origin.

The equation for the line is $y = 48x - 12$.

(a) (i) During one test there is a constant windspeed of 5·8 m s^{-1}. Show that the angular velocity of the anemometer at this windspeed is 28 rad s^{-1}.   **1**

    (ii) In a **second** test the windspeed is reduced from 5·8 m s^{-1} to 1·6 m s^{-1} in a time of 8·0 seconds. Calculate the angular acceleration of the anemometer.   **3**

Marks

2. (continued)

(b) The rotating part of the anemometer is made up of a central cylinder and three arms as shown in Figure 2C.

Plan view of rotating parts of anemometer

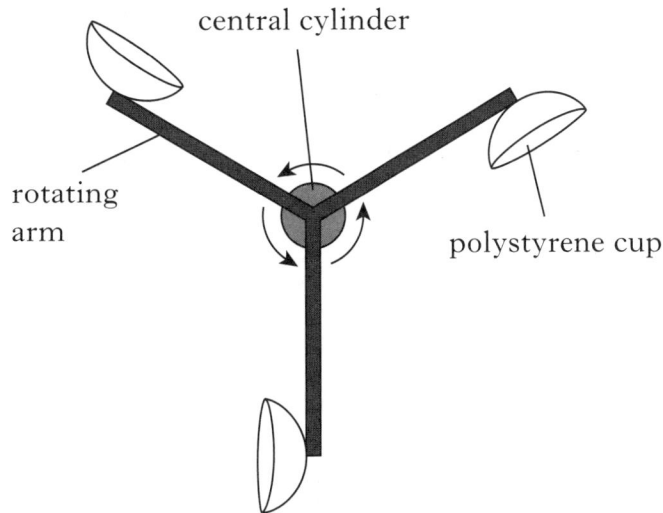

Figure 2C

Each arm consists of a rod of length 76 mm and mass 11 g with a polystyrene cup of negligible mass on the end of the rod. One arm of the anemometer is shown in Figure 2D.

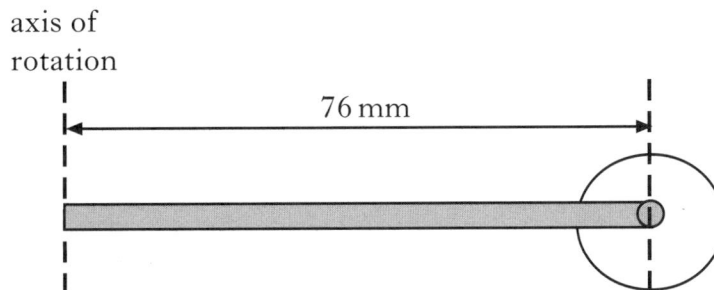

Figure 2D

(i) Calculate the moment of inertia of the rod about the axis shown.   2

(ii) The moment of inertia of the central cylinder is $1 \cdot 1 \times 10^{-6} \, \text{kg m}^2$. Calculate the total moment of inertia of all the rotating parts.   2

(c) Calculate the average torque acting on the rotating system during the **second** test.   2

(d) The anemometer is set up at a field site to record wind speed. Freezing rain deposits a layer of ice evenly over the cups. Explain the effect this might have on the motion of the arm.   2

(12)

Marks

3. An electric toothbrush has a circular brush head, diameter 12 mm, as shown in Figure 3A.

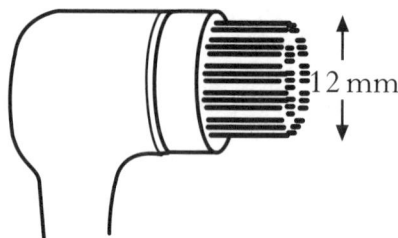

Figure 3A

The toothbrush has two settings.

On setting 1, the brush head vibrates with simple harmonic motion. On this setting, the head vibrates with a frequency of 33 Hz and moves a maximum distance of 4·2 mm as shown in Figure 3B.

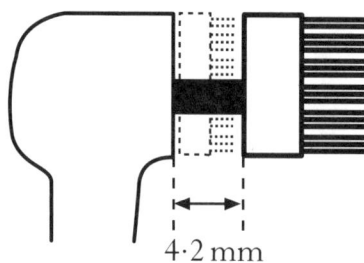

Figure 3B

  (*a*) (i) Write an expression for the displacement, in metres, of the brush head. **2**

       (ii) Calculate the maximum speed of the brush head. **2**

Marks

3. (continued)

(b) On setting 2, the brush head can be considered to oscillate with simple harmonic motion with amplitude A as shown in Figure 3C.

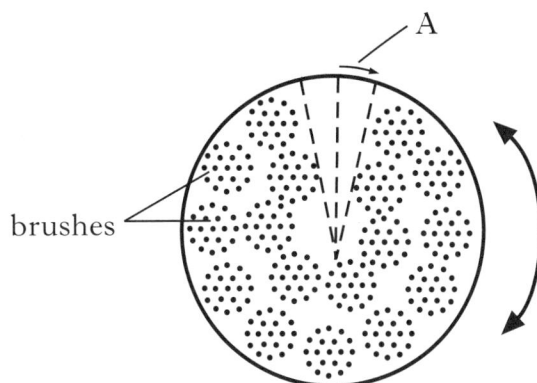

Figure 3C

The velocity, in $m\,s^{-1}$, of a point on the circumference of the head can be given by the equation

$$v = 9 \cdot 2 \times 10^{-2} \cos 77t. \; \ast$$

Calculate the amplitude of the oscillation on setting 2.

2

(c) A particle of toothpaste of mass $2 \cdot 5 \times 10^{-6}\,kg$ on the outer edge of the brush head is shown in Figure 3D.

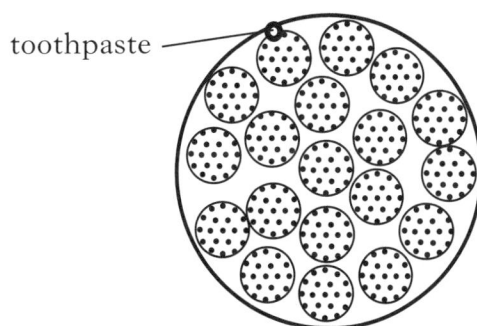

Figure 3D

The switch is on setting 2.

(i) Calculate the maximum kinetic energy of the particle of toothpaste.

2

(ii) Sketch a graph of the kinetic energy of the particle of toothpaste against its displacement. Appropriate numerical values are required on both axes.

2

(10)

[Turn over

Marks

4. (*a*) A point charge of $+4 \cdot 0 \, \mu C$ is shown in Figure 4A.

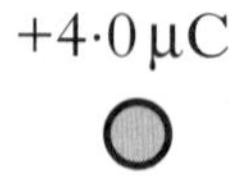

$+4 \cdot 0 \, \mu C$

Figure 4A

(i) Copy Figure 4A and draw the electric field lines around this point charge.

1

(ii) A point charge of $-2 \cdot 0 \, \mu C$ is now placed at a distance of $0 \cdot 10 \, m$ from the first charge as shown in Figure 4B.

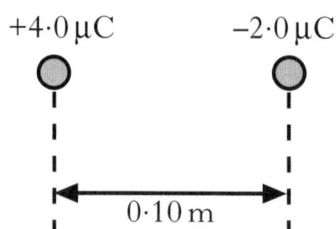

$+4 \cdot 0 \, \mu C$ $-2 \cdot 0 \, \mu C$

$0 \cdot 10 \, m$

Figure 4B

Explain why the electric field strength **is not** zero at any point between these two charges.

2

(iii) Point P is $0 \cdot 24 \, m$ to the right of the second charge as shown in Figure 4C.

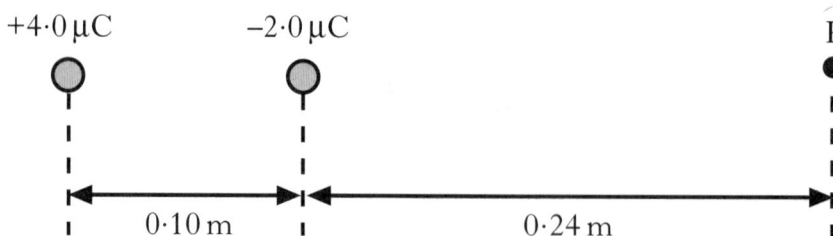

$+4 \cdot 0 \, \mu C$ $-2 \cdot 0 \, \mu C$ P

$0 \cdot 10 \, m$ $0 \cdot 24 \, m$

Figure 4C

Calculate the electric field strength at point P.

3

(*b*) Two like charges experience a repulsive electrostatic force.

Explain why two protons in a nucleus do not fly apart.

2

Marks

4. (continued)

(c) Information on the properties of three elementary particles together with two types of quarks and their corresponding antiquarks is shown in Figure 4D.

Properties of elementary particles			
Particle	Number of quarks	Charge	Baryon number
Proton	3	$+e$	1
Antiproton	3	$-e$	-1
Pi-meson	2	$-e$	0

Properties of quarks and antiquarks		
Particle	Charge	Baryon number
Up quark	$+\frac{2}{3}e$	$+\frac{1}{3}$
Down quark	$-\frac{1}{3}e$	$+\frac{1}{3}$
Anti-up quark	$-\frac{2}{3}e$	$-\frac{1}{3}$
Anti-down quark	$+\frac{1}{3}e$	$-\frac{1}{3}$

Figure 4D

(i) Using information from Figure 4D, show that a proton consists of two up quarks and one down quark.   1

(ii) State the combination of quarks that forms a pi-meson.   1

(10)

[Turn over

Marks

5. An alpha particle passes through a region that has perpendicular electric and magnetic fields, as shown in Figure 5A.

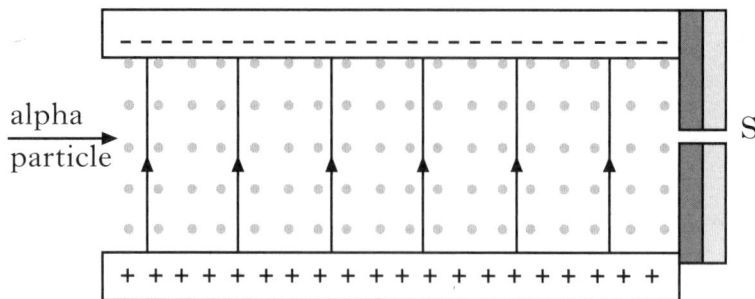

Figure 5A

The magnetic induction is 6·8 T and is directed out of the page.

The force on the alpha particle due to the magnetic field is $5 \cdot 0 \times 10^{-11}$ N.

(a) Show that the velocity of the alpha particle is $2 \cdot 3 \times 10^{7}\,\mathrm{m\,s^{-1}}$.   **2**

(b) In order that the alpha particle exits through slit S, it must pass through the region undeflected.

Calculate the strength of the electric field that ensures the alpha particle exits through slit S.   **2**

Marks

5. **(continued)**

(c) After passing through slit S, the alpha particle enters a region where there is the same uniform perpendicular magnetic field but no electric field as shown in Figure 5B.

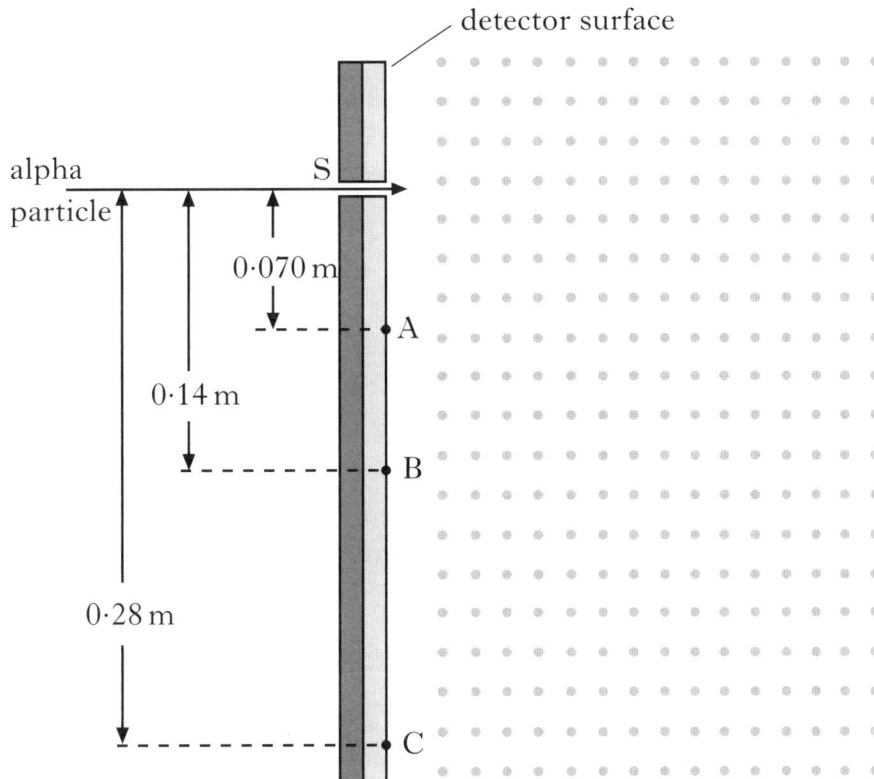

Figure 5B

This magnetic field causes the alpha particle to travel in a semi-circular path and hit the detector surface.

Points A, B and C are at distances of 0·070 m, 0·14 m and 0·28 m respectively from slit S.

Show, by calculation, which point on the detector surface is hit by the alpha particle.   3

(d) An electron travelling at the same speed as the alpha particle passes through slit S into the region of uniform magnetic field.

State **two** differences in the semi-circular path of the electron compared to the path of the alpha particle. **Justify** your answer.   3

(10)

[Turn over

Marks

6. (*a*) A 3·0 V battery is connected in series with a switch, a resistor and an inductor of negligible resistance. A neon lamp is connected across the inductor as shown in Figure 6A.

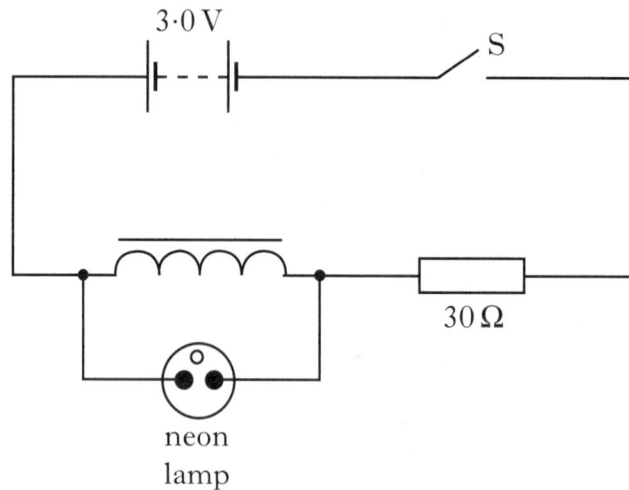

Figure 6A

(i) Sketch a graph to show how the current in the inductor varies with time from the instant the switch is closed.

Appropriate numerical values are required on the current axis. **2**

(ii) The neon lamp requires a potential difference of at least 110 V across it before it lights.

Explain why the lamp does not light when the switch is closed. **1**

(iii) After a few seconds the switch is opened and the lamp flashes.

Explain, in terms of the magnetic field, why the lamp flashes as the switch is opened. **2**

(iv) The neon lamp has an average power of 1·2 mW and a flash that lasts 0·25 s.

Assuming all the energy stored by the inductor is transferred to the lamp, calculate the inductance of the inductor. **3**

Marks

6. (continued)

(b) Figure 6B shows a circuit used to investigate the relationship between the current in an inductive circuit and the supply frequency.

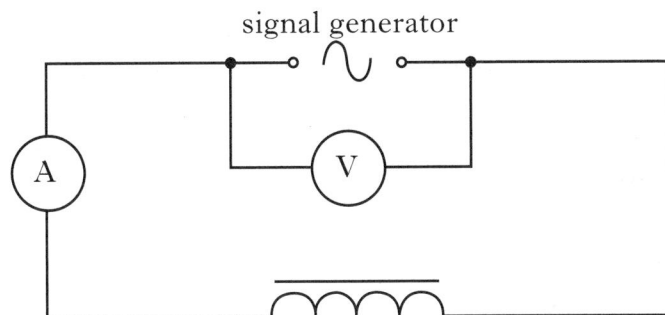

Figure 6B

The reading on the ammeter is noted for different values of supply frequency.

(i) State the purpose of the voltmeter. 1

(ii) Describe how the data obtained should be analysed to determine the relationship between the current in the inductive circuit and the supply frequency. 1

(iii) State the expected relationship. 1

(c) A loudspeaker system is connected to a music amplifier. The system contains a capacitor, inductor and two loudspeakers, LS1 and LS2, as shown in Figure 6C.

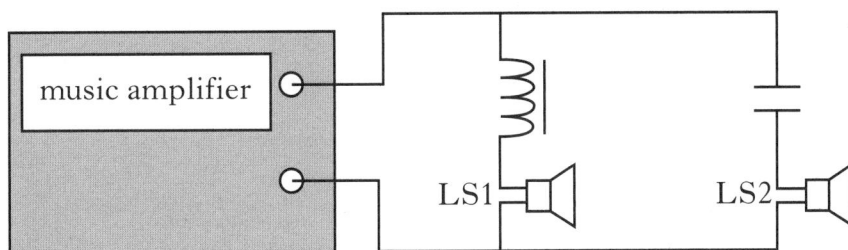

Figure 6C

The circuit is designed so that one loudspeaker emits low frequency sounds while the other emits high frequency sounds.

By comparing the capacitive and inductive reactances, describe the operation of this system. 2

(13)

[Turn over

Marks

7. The Bohr model of the hydrogen atom states that the angular momentum of an electron is quantised.

(a) (i) Calculate the minimum angular momentum of the electron in a hydrogen atom. 2

(ii) When the electron has its minimum angular momentum it is in an orbit of radius $5 \cdot 3 \times 10^{-11}$ m. Calculate the linear momentum of the electron in this orbit. 2

(iii) Calculate the de Broglie wavelength of the electron in this orbit. 2

(b) One of the limitations of the Bohr model is that an orbiting electron is constantly accelerating and therefore should continuously emit electromagnetic radiation.

(i) What would happen to the orbit of the electron if electromagnetic radiation were to be continuously emitted? 1

(ii) What is the name of the branch of physics that provides methods to determine the electron's position in terms of probability? 1

(8)

Marks

8. A student uses a probe to measure the magnetic induction near a long straight current carrying conductor PQ, as shown in Figure 8A.

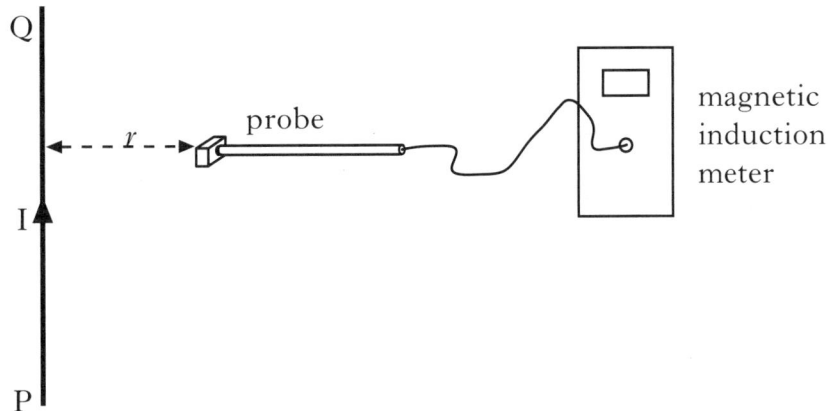

Figure 8A

The following data is obtained.

Distance r/m	Magnetic induction B/T
0·25	$1·7 \times 10^{-7}$

(a) Calculate the current in the conductor PQ.   **2**

(b) The unit of magnetic induction is the tesla. Define *one tesla*.   **1**

(c) A second long straight conductor RS carrying a current of 2 A is placed at a distance of 0·25 m from the first conductor PQ, as shown in Figure 8B.

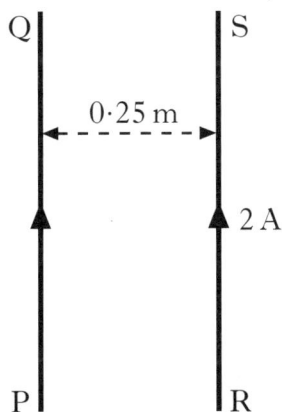

Figure 8B

Calculate the magnitude of the force per metre acting on conductor RS.   **2**

(5)

[Turn over

Marks

9. (*a*) A student is measuring the thickness of a piece of paper using a thin air wedge. The air wedge is formed between two glass plates that are in contact at one end and separated by a sheet of paper at the other end.

Monochromatic light is reflected down onto the air wedge. A travelling microscope is used to view the resulting interference pattern as shown in Figure 9A.

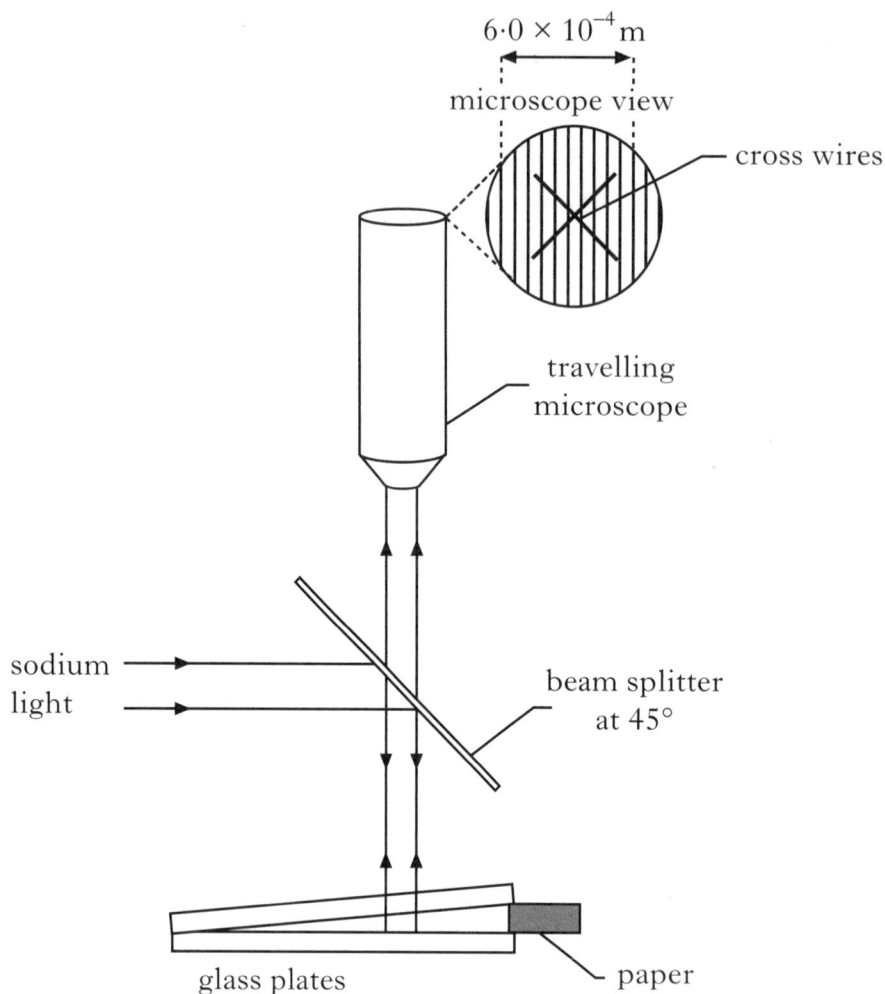

Figure 9A

   (i) The air wedge produces interference by division of amplitude.

     State what is meant by the term *division of amplitude*.   **1**

  (ii) The following data is obtained.

       10 fringe separations = $(6{\cdot}0 \pm 0{\cdot}5) \times 10^{-4}\,$m
          Wavelength, $\lambda = 580\,$nm $\pm\, 10\,$nm
     Length of glass plate, $l = (4{\cdot}0 \pm 0{\cdot}1) \times 10^{-2}\,$m

     Calculate the thickness of the paper.   **3**

 (iii) Calculate the percentage uncertainty in the thickness.   **2**

Marks

9. (continued)

(b) A beam of monochromatic light of wavelength 580 nm illuminates a film of soap that is held in a wire loop. An interference pattern is produced as shown in Figure 9B.

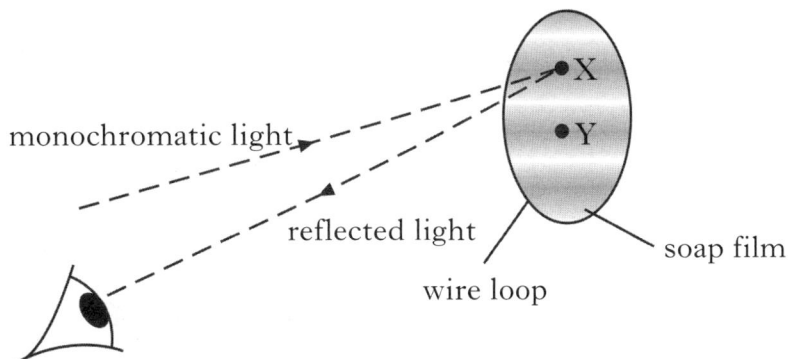

Figure 9B

An expanded view of part of the film is shown in Figure 9C.

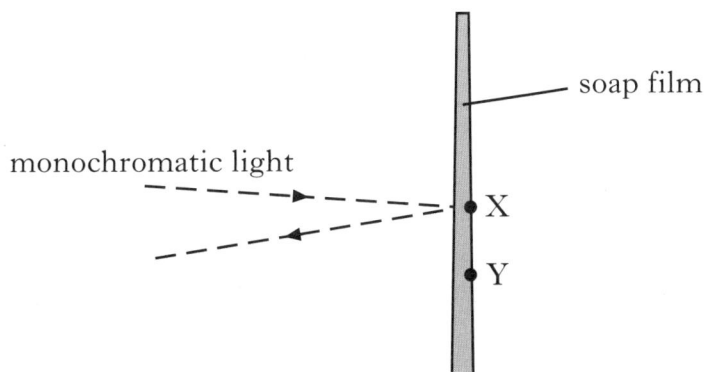

Figure 9C

(i) Destructive interference occurs when light is reflected at position X.

Explain why destructive interference occurs. You may include a diagram as part of your answer.   2

(ii) At position X the thickness of the film is 4·00 μm.

The refractive index of the film of soap is 1·45.

Calculate the optical path difference produced by this film at position X.   2

(iii) The **next** position where destructive interference occurs is position Y where the film is slightly thicker.

Calculate the optical path difference produced by the film at position Y.   2

(12)

Marks

10. A long pipe containing polystyrene beads is closed by a plunger. A loudspeaker at the other end is connected to a signal generator as shown in Figure 10A.

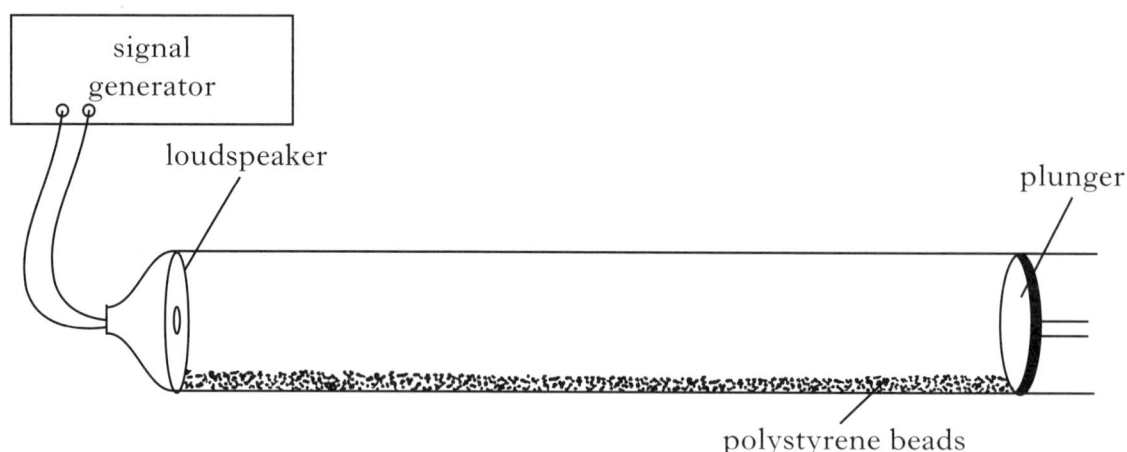

Figure 10A

The loudspeaker is switched on and the frequency is adjusted until a stationary sound wave is set up in the tube. The polystyrene beads form the pattern shown in Figure 10B. The distance between successive piles of beads is 85 mm.

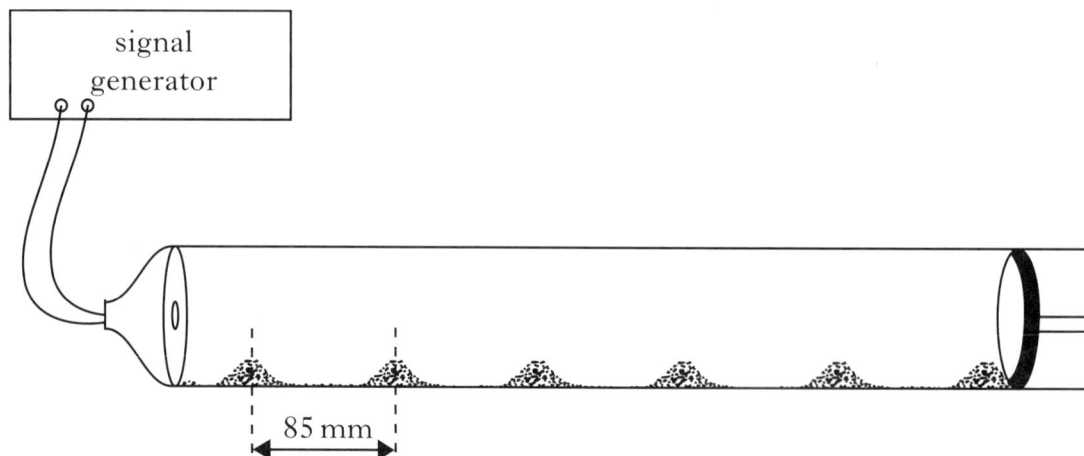

Figure 10B

(a) Explain how sound waves in the tube produce the stationary wave.   1

(b) Define the terms node and antinode.   1

(c) Do the polystyrene beads collect at nodes or antinodes? Justify your answer.   2

(d) The signal generator is set to 1·95 kHz. Calculate the speed of sound in air.   3

(7)

[END OF QUESTION PAPER]

ADVANCED HIGHER

2010

[BLANK PAGE]

X069/701

NATIONAL QUALIFICATIONS 2010	FRIDAY, 28 MAY 1.00 PM – 3.30 PM	PHYSICS ADVANCED HIGHER

Reference may be made to the Physics Data Booklet.

Answer **all** questions.

Any necessary data may be found in the Data Sheet on page two.

Care should be taken to give an appropriate number of significant figures in the final answers to calculations.

Square-ruled paper (if used) should be placed inside the front cover of the answer book for return to the Scottish Qualifications Authority.

SQA

DATA SHEET

COMMON PHYSICAL QUANTITIES

Quantity	Symbol	Value	Quantity	Symbol	Value
Gravitational acceleration on Earth	g	$9 \cdot 8 \text{ m s}^{-2}$	Mass of electron	m_e	$9 \cdot 11 \times 10^{-31} \text{ kg}$
Radius of Earth	R_E	$6 \cdot 4 \times 10^6 \text{ m}$	Charge on electron	e	$-1 \cdot 60 \times 10^{-19} \text{ C}$
Mass of Earth	M_E	$6 \cdot 0 \times 10^{24} \text{ kg}$	Mass of neutron	m_n	$1 \cdot 675 \times 10^{-27} \text{ kg}$
Mass of Moon	M_M	$7 \cdot 3 \times 10^{22} \text{ kg}$	Mass of proton	m_p	$1 \cdot 673 \times 10^{-27} \text{ kg}$
Radius of Moon	R_M	$1 \cdot 7 \times 10^6 \text{ m}$	Mass of alpha particle	m_α	$6 \cdot 645 \times 10^{-27} \text{ kg}$
Mean Radius of Moon Orbit		$3 \cdot 84 \times 10^8 \text{ m}$	Charge on alpha particle		$3 \cdot 20 \times 10^{-19} \text{ C}$
Universal constant of gravitation	G	$6 \cdot 67 \times 10^{-11} \text{ m}^3 \text{ kg}^{-1} \text{ s}^{-2}$	Planck's constant	h	$6 \cdot 63 \times 10^{-34} \text{ J s}$
Speed of light in vacuum	c	$3 \cdot 0 \times 10^8 \text{ m s}^{-1}$	Permittivity of free space	ε_0	$8 \cdot 85 \times 10^{-12} \text{ F m}^{-1}$
Speed of sound in air	v	$3 \cdot 4 \times 10^2 \text{ m s}^{-1}$	Permeability of free space	μ_0	$4\pi \times 10^{-7} \text{ H m}^{-1}$

REFRACTIVE INDICES

The refractive indices refer to sodium light of wavelength 589 nm and to substances at a temperature of 273 K.

Substance	Refractive index	Substance	Refractive index
Diamond	2·42	Glycerol	1·47
Glass	1·51	Water	1·33
Ice	1·31	Air	1·00
Perspex	1·49	Magnesium Fluoride	1·38

SPECTRAL LINES

Element	Wavelength/nm	Colour	Element	Wavelength/nm	Colour
Hydrogen	656	Red	Cadmium	644	Red
	486	Blue-green		509	Green
	434	Blue-violet		480	Blue
	410	Violet		*Lasers*	
	397	Ultraviolet	Element	Wavelength/nm	Colour
	389	Ultraviolet			
			Carbon dioxide	9550 } 10590 }	Infrared
Sodium	589	Yellow	Helium-neon	633	Red

PROPERTIES OF SELECTED MATERIALS

Substance	Density/ kg m^{-3}	Melting Point/ K	Boiling Point/ K	Specific Heat Capacity/ $\text{J kg}^{-1} \text{ K}^{-1}$	Specific Latent Heat of Fusion/ J kg^{-1}	Specific Latent Heat of Vaporisation/ J kg^{-1}
Aluminium	$2 \cdot 70 \times 10^3$	933	2623	$9 \cdot 02 \times 10^2$	$3 \cdot 95 \times 10^5$	. . . .
Copper	$8 \cdot 96 \times 10^3$	1357	2853	$3 \cdot 86 \times 10^2$	$2 \cdot 05 \times 10^5$	. . . .
Glass	$2 \cdot 60 \times 10^3$	1400	. . . .	$6 \cdot 70 \times 10^2$	. . . .	. . . .
Ice	$9 \cdot 20 \times 10^2$	273	. . . .	$2 \cdot 10 \times 10^3$	$3 \cdot 34 \times 10^5$	. . . .
Glycerol	$1 \cdot 26 \times 10^3$	291	563	$2 \cdot 43 \times 10^3$	$1 \cdot 81 \times 10^5$	$8 \cdot 30 \times 10^5$
Methanol	$7 \cdot 91 \times 10^2$	175	338	$2 \cdot 52 \times 10^3$	$9 \cdot 9 \times 10^4$	$1 \cdot 12 \times 10^6$
Sea Water	$1 \cdot 02 \times 10^3$	264	377	$3 \cdot 93 \times 10^3$	. . . .	. . . .
Water	$1 \cdot 00 \times 10^3$	273	373	$4 \cdot 19 \times 10^3$	$3 \cdot 34 \times 10^5$	$2 \cdot 26 \times 10^6$
Air	1·29	. . . .	. . . .			
Hydrogen	$9 \cdot 0 \times 10^{-2}$	14	20	$1 \cdot 43 \times 10^4$	. . . .	$4 \cdot 50 \times 10^5$
Nitrogen	1·25	63	77	$1 \cdot 04 \times 10^3$	. . . .	$2 \cdot 00 \times 10^5$
Oxygen	1·43	55	90	$9 \cdot 18 \times 10^2$	. . . .	$2 \cdot 40 \times 10^5$

The gas densities refer to a temperature of 273 K and a pressure of $1 \cdot 01 \times 10^5$ Pa.

[BLANK PAGE]

Marks

1. A turntable, radius r, rotates with a constant angular velocity ω about an axis of rotation. Point X on the circumference of the turntable is moving with a tangential speed v, as shown in Figure 1A.

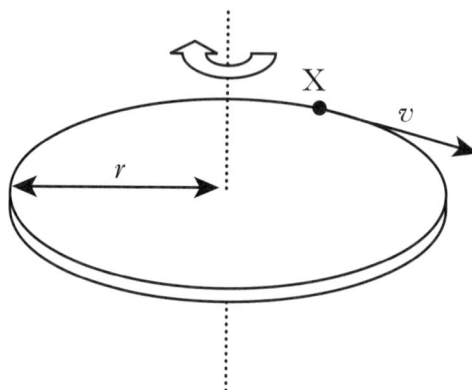

Figure 1A

(a) Derive the relationship

$$v = r\,w.$$

2

(b) Data recorded for the turntable is shown below.

Angle of rotation	$(3\cdot1 \pm 0\cdot1)$ rad
Time taken for angle of rotation	$(4\cdot5 \pm 0\cdot1)$ s
Radius of disk	$(0\cdot148 \pm 0\cdot001)$ m

   (i) Calculate the tangential speed v.   2

  (ii) Calculate the percentage uncertainty in this value of v.   2

 (iii) As the disk rotates, v remains constant.

     (A) Explain why point X is accelerating.   1

     (B) State the direction of this acceleration.   1

(8)

Marks

2. A motorised model plane is attached to a light string anchored to a ceiling.

   The plane follows a circular path of radius 0·35 m as shown in Figure 2A.

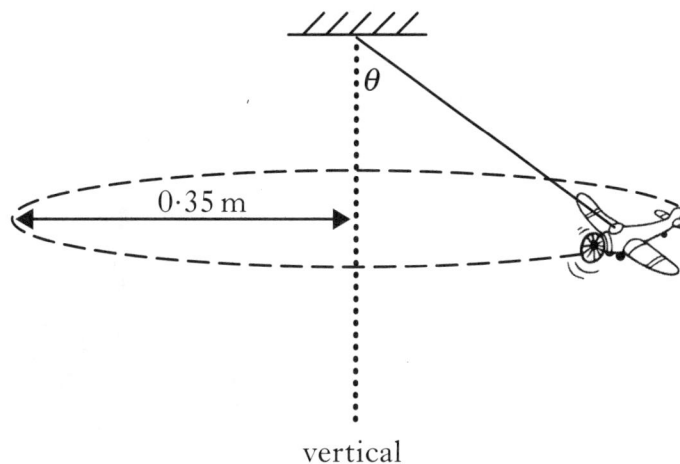

vertical

Figure 2A

The plane has a mass of 0·20 kg and moves with a constant angular velocity of 6·0 rad s^{-1}.

(a) Calculate the central force acting on the plane.   2

(b) Calculate angle θ of the string to the vertical.   2

(c) What effect would a decrease in the plane's speed have on angle θ?

   Justify your answer.   2

                                                                                      (6)

[Turn over

3. A mass of 2·5 kg is attached to a string of negligible mass. The string is wound round a flywheel of radius 0·14 m. A motion sensor, connected to a computer, is placed below the mass as shown in Figure 3A.

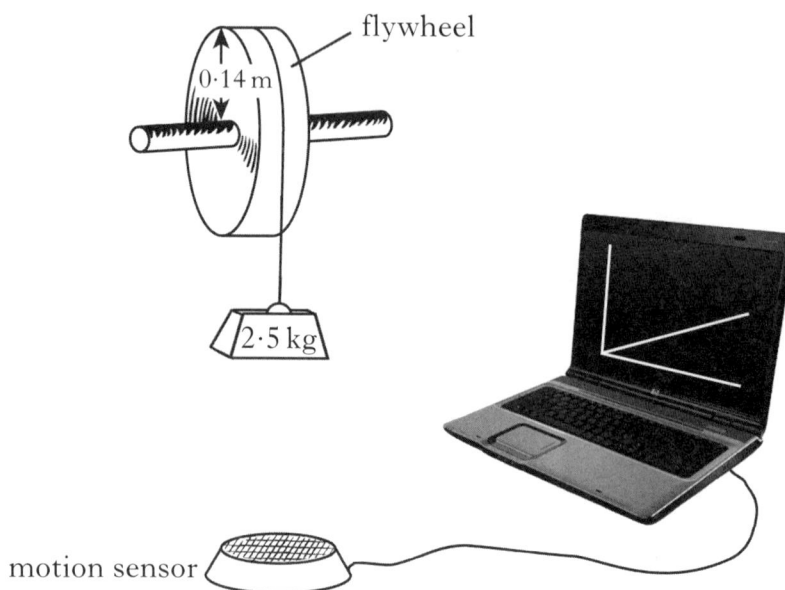

Figure 3A

The mass is released from rest. The computer calculates the linear velocity of the mass as it falls and the angular velocity of the flywheel.

The graph of the angular velocity of the flywheel against time, as displayed on the computer, is shown in Figure 3B.

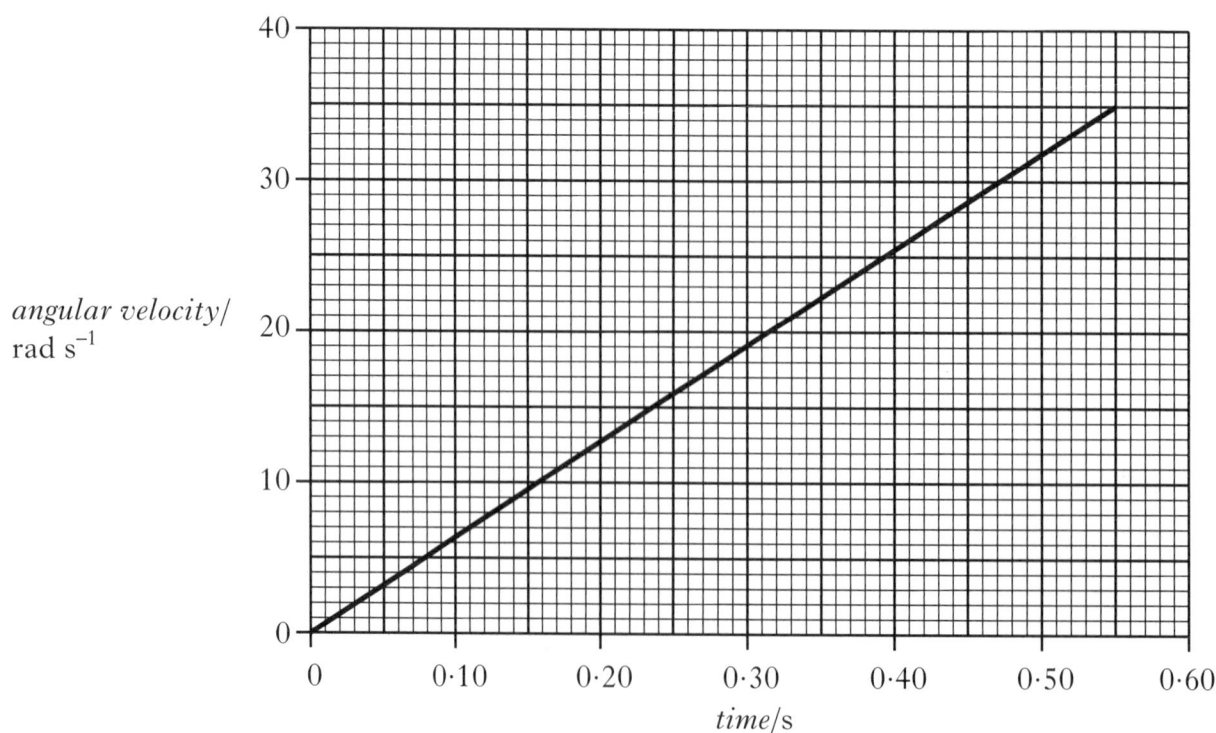

Figure 3B

Marks

3. (continued)

(a) Calculate the angular acceleration of the flywheel. **2**

(b) Show that the mass falls a distance of $1 \cdot 3$ m in the first $0 \cdot 55$ seconds. **3**

(c) Use the conservation of energy to calculate the moment of inertia of the flywheel. Assume the frictional force to be negligible. **4**

(9)

[Turn over

Marks

4. (a) State the law of conservation of angular momentum.   **1**

(b) A student sits on a platform that is free to rotate on a frictionless bearing. The angular velocity of the rotating platform is displayed on a computer.

The student rotates with a hand outstretched at 0·60 m from the axis of rotation as shown in Figure 4A.

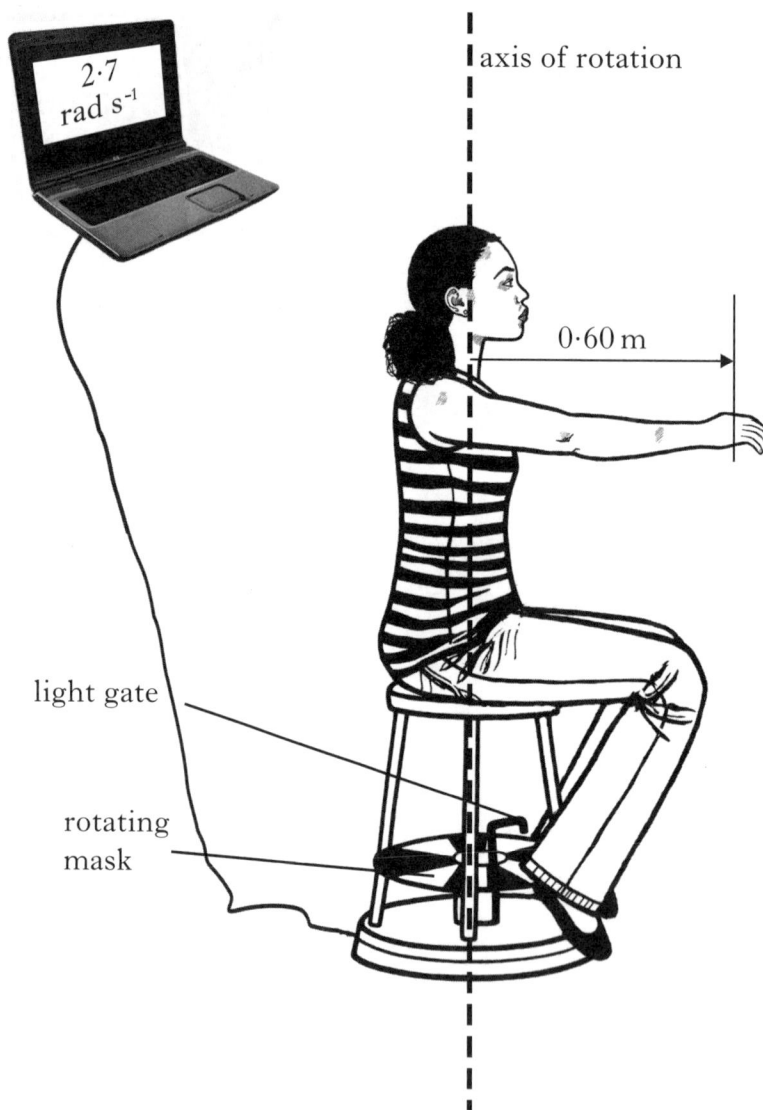

Figure 4A

The moment of inertia of the student and rotating platform is 4·1 kg m². The angular velocity is 2·7 rad s⁻¹.

(i) Calculate the angular momentum of the student and rotating platform.   **2**

4. (b) (continued)   *Marks*

(ii) As the student rotates, she grasps a 2·5 kg mass from a stand as shown in Figure 4B.

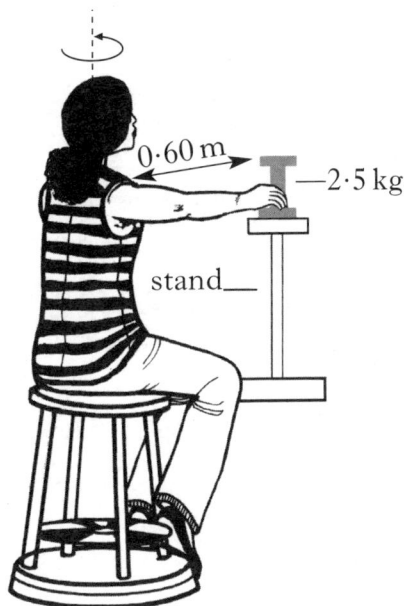

Figure 4B

Calculate the angular velocity of the student and platform just after the mass has been grasped.   3

(iii) The student then pulls the mass towards her body.

Explain the effect this has on the angular velocity of the student and the platform.   2

(c) In another investigation the student and platform rotate at 1·5 rad s^{-1}. The student puts one foot on the floor as shown in Figure 4C.

Figure 4C

The frictional force between the student's shoe and the floor brings the student and platform to rest in 0·75 seconds. The new moment of inertia of the student and platform is 4·5 kg m^{2}.

Calculate the average frictional torque.   3

                                                                        (11)

5. A motorised mixer in a DIY store is used to mix different coloured paints.

Paints are placed in a tin and the tin is clamped to the base as shown in Figure 5A.

Figure 5A

The oscillation of the tin in the vertical plane closely approximates to simple harmonic motion.

The amplitude of the oscillation is 0·012 m.

The mass of the tin of paint is 1·4 kg.

Figure 5B shows the graph of the acceleration against displacement for the tin of paint.

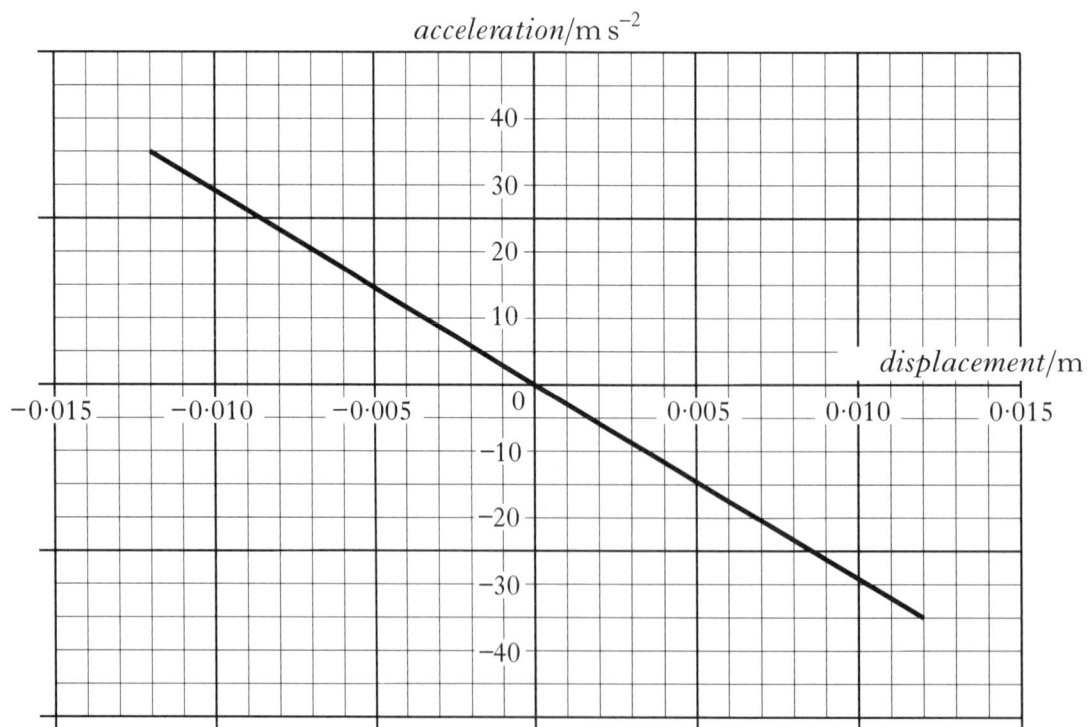

Figure 5B

Marks

5. **(continued)**

   (*a*) Show that the angular frequency ω of the oscillation is $54\,\text{rad s}^{-1}$.   **1**

   (*b*) Write an expression for the displacement *y* of the tin as a function of time. Include appropriate numerical values.   **2**

   (*c*) Derive an expression for the velocity *v* of the tin as a function of time. Numerical values should again be included.   **2**

   (*d*) Calculate the maximum kinetic energy of the tin of paint as it oscillates.   **2**

   (7)

[Turn over

Marks

6. A hollow metal sphere, radius $1 \cdot 00$ mm, carries a charge of $-1 \cdot 92 \times 10^{-12}$ C.

   (a) Calculate the electric field strength, E, at the surface of the sphere.   **2**

   (b) Four students sketch graphs of the variation of electric field strength with distance from the centre of the sphere as shown in Figure 6A.

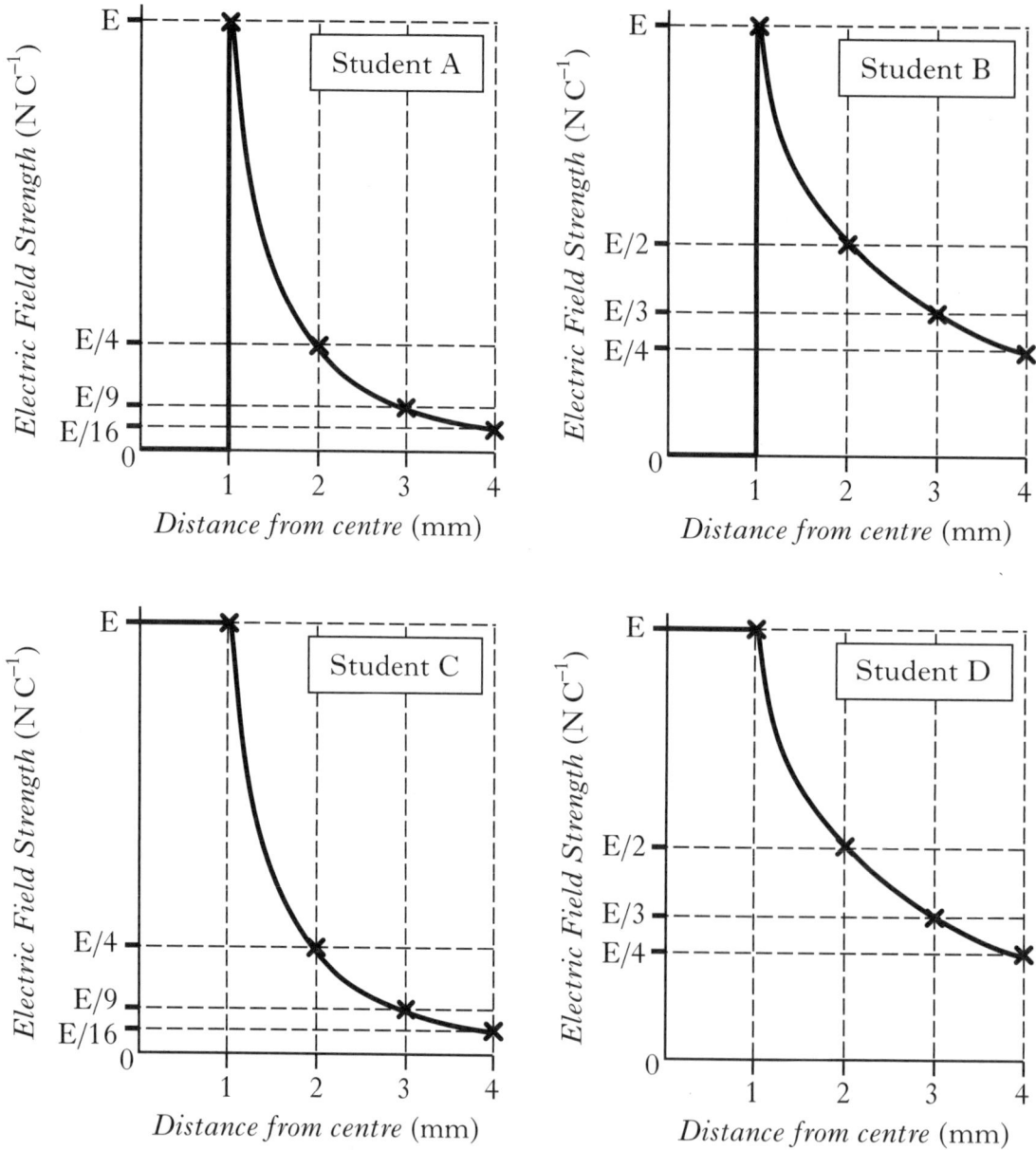

Figure 6A

   (i) Which student has drawn the correct graph?   **1**

   (ii) Give **two** reasons to support your choice.   **2**

Marks

6. (continued)

(c) Four point charges, Q_1, Q_2, Q_3 and Q_4, each of value $-2 \cdot 97 \times 10^{-8}$ C, are held in a square array. The hollow sphere with charge $-1 \cdot 92 \times 10^{-12}$ C is placed $30 \cdot 0$ mm above the centre of the array where it is held stationary by an electrostatic force.
The hollow sphere is $41 \cdot 2$ mm from each of the four charges as shown in Figure 6B.

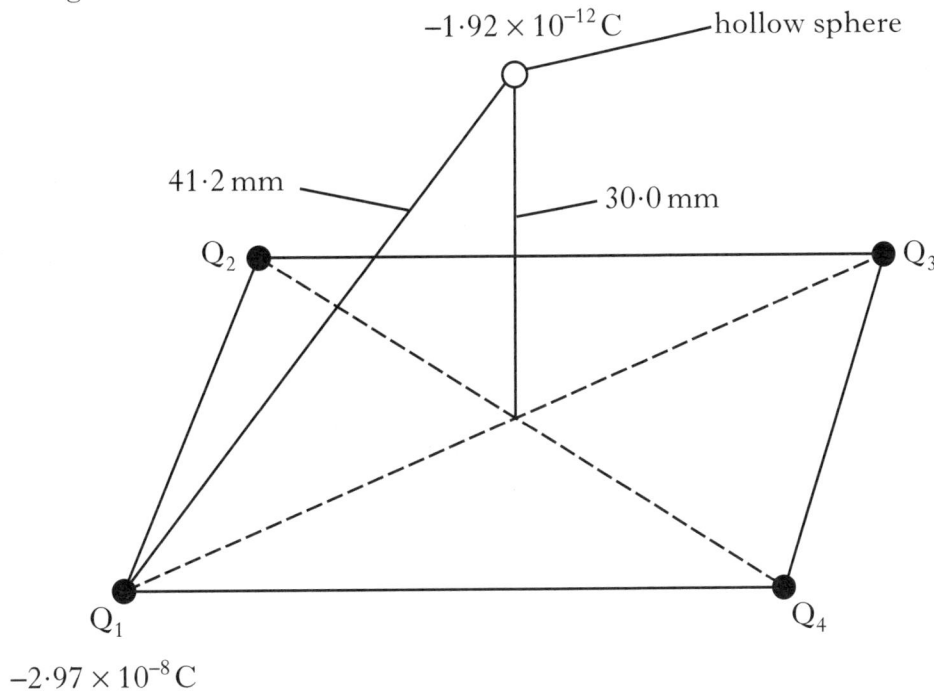

Figure 6B

(i) Calculate the magnitude of the force acting on the sphere due to charge Q_1.   2

(ii) Calculate the vertical component of this force.   2

(iii) Calculate the resultant electrostatic force on the sphere due to the whole array.   1

(iv) Calculate the mass of the sphere.   2

    (12)

[Turn over

Marks

7. A beam of protons enters a region of uniform magnetic field, at right angles to the field.

   The protons follow a circular path in the magnetic field until a potential difference is applied across the deflecting plates. The deflected protons hit a copper target. The protons travel through a vacuum. A simplified diagram is shown in Figure 7A.

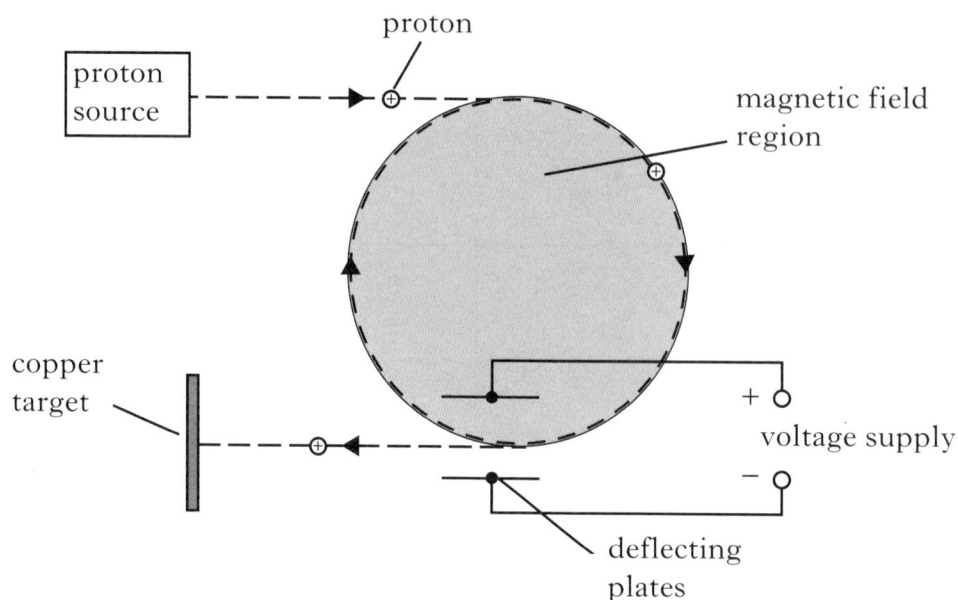

Figure 7A

   (a) State the direction of the magnetic field, B.   **1**

   (b) The speed of the protons is $6 \cdot 0 \times 10^6\,\mathrm{m\,s^{-1}}$ and the magnetic induction is $0 \cdot 75\,\mathrm{T}$. Calculate the radius of the circular path followed by the protons.   **3**

   (c) Calculate the electric field strength required to make the protons move off at a tangent to the circle.   **2**

Marks

7. (continued)

(d) A proton of charge q initially travels at speed v directly towards a copper nucleus as shown in Figure 7B. The copper nucleus has charge Q.

copper nucleus

proton

v

Figure 7B

(i) Show that the distance of closest approach, r, to the copper nucleus is given by

$$\frac{qQ}{2\pi\varepsilon_0 mv^2}.$$

1

(ii) Calculate the distance of closest approach for a proton initially travelling at $6.0 \times 10^6\,\mathrm{m\,s^{-1}}$.

3

(iii) Name the force that holds the protons together in the copper nucleus.

1

(e) The beam of protons in Figure 7A is replaced by a beam of electrons. The speed of the electrons is the same as the speed of the protons.

State **two** changes that must be made to the magnetic field to allow the electrons to follow the same circular path as the protons.

2

(13)

[Turn over

Marks

8. Identification of elements in a semiconductor sample can be carried out using an electron scanner to release atoms from the surface of the sample for analysis. Electrons are accelerated from rest between a cathode and anode by a potential difference of 2·40 kV.

A variable voltage supply connected to the deflection plates enables the beam to scan the sample between points A and B shown in Figure 8A.

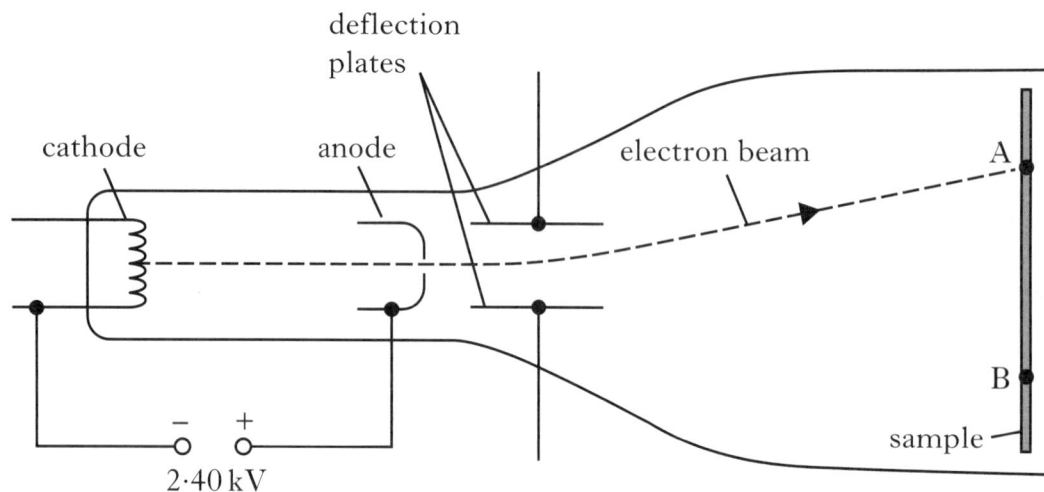

Figure 8A

(a) Calculate the speed of the electrons as they pass through the anode. **2**

(b) Explain why the electron beam follows

  (i) a curved path between the plates; **1**

  (ii) a straight path beyond the plates. **1**

When the potential difference across the deflection plates is 100 V, the electron beam strikes the sample at position A.

(c) The deflection plates are 15·0 mm long and separated by 10·0 mm.

  (i) Show that the vertical acceleration between the plates is $1·76 \times 10^{15} \text{m s}^{-2}$. **2**

  (ii) Calculate the vertical velocity of an electron as it emerges from between the plates. **3**

(d) The anode voltage is now increased. State what happens to the length of the sample scanned by the electron beam.

  You must justify your answer. **2**

(11)

Marks

9. A transverse wave travels along a string as shown in Figure 9A.

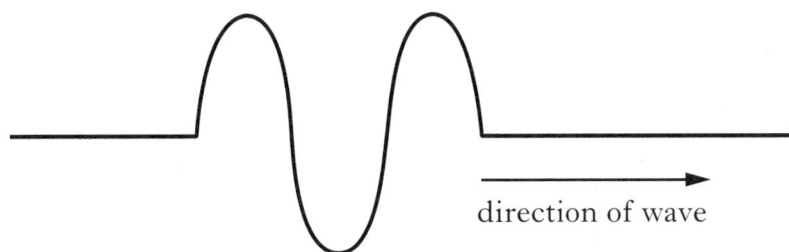

Figure 9A

The equation representing the travelling wave on the string is

$$y = 8 \cdot 6 \times 10^{-2} \sin 2\pi \, (2 \cdot 4t - 2 \cdot 0x)$$

where x and y are in metres and t is in seconds.

(a) State the frequency of the wave.   1

(b) Calculate the velocity of the wave.   2

(c) Attached to the end of the string is a very light ring. The ring is free to move up and down a fixed vertical rod.

Figure 9B shows the string after the wave reflects from the vertical rod.

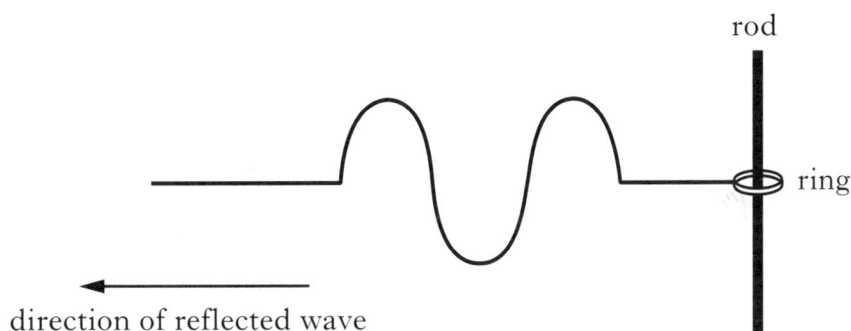

Figure 9B

When the wave reflects, its intensity falls to one quarter of its original value. The frequency and wavelength are constant.

Write the equation that represents this reflected wave.   2

(5)

[Turn over

Marks

10. (a) Explain the formation of coloured fringes when white light illuminates a thin film of oil on a water surface.

2

(b) Thin film coatings deposited on glass can be used to make the glass non-reflecting for certain wavelengths of light, as shown in Figure 10A.

The refractive index of the coating is less than glass, but greater than air.

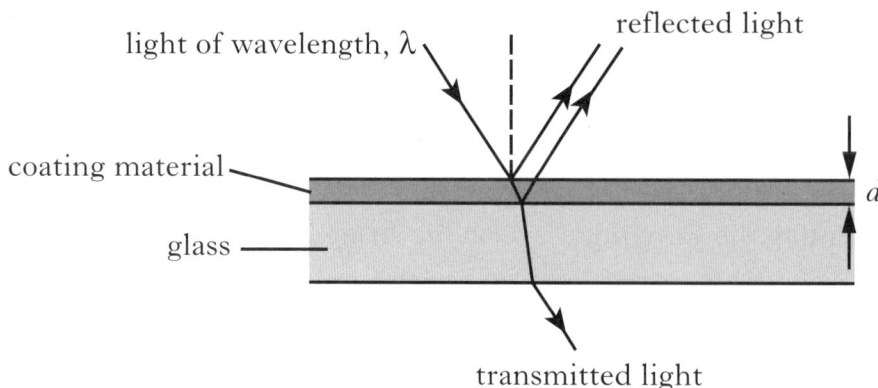

light of wavelength, λ reflected light

coating material

glass

d

transmitted light

Figure 10A

Show that for near normal incidence

$$d = \frac{\lambda}{4n}$$

where n is the refractive index of the coating material and d is the thinnest coating that will be non-reflecting for light of wavelength, λ.

2

(c) The relationship in (b) also applies to radiation of wavelength 780 nm.

A thin film coating has a refractive index of 1·30. For radiation of wavelength 780 nm the minimum thickness for a thin film that is non-reflecting is 0·150 μm. In practice, this thickness is too thin to manufacture.

Calculate the thickness of the next thinnest coating that would be non-reflecting for this wavelength.

2

Marks

10. (continued)

(*d*) Six laser beams provide photons of wavelength 780 nm. These photons collide with rubidium atoms and cause the atoms to come to rest, as shown in Figure 10B.

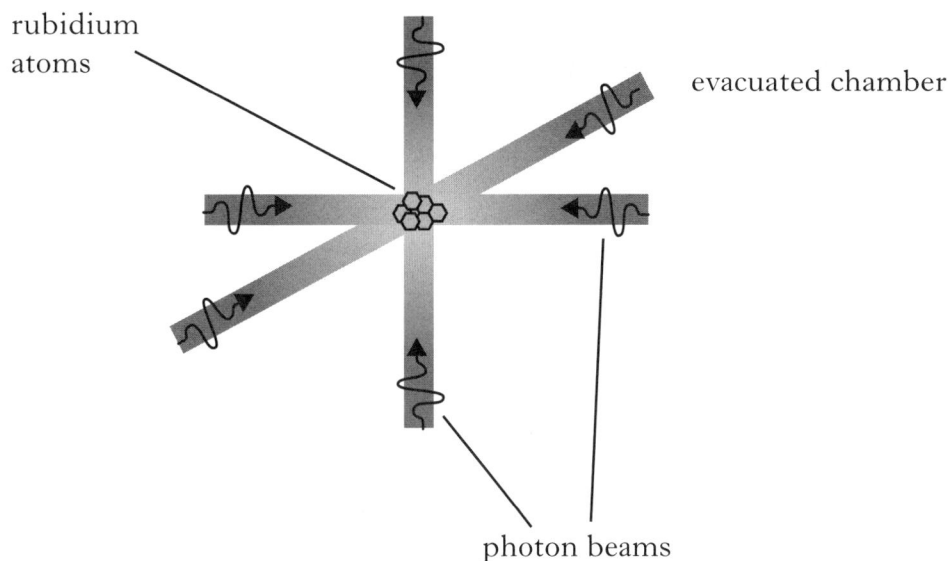

Figure 10B

Each rubidium atom has a mass of 1.43×10^{-25} kg and kinetic energy of 4.12×10^{-21} J before the lasers are switched on.

(i) Calculate the momentum of a rubidium atom before the lasers are switched on. **3**

(ii) Calculate the momentum of each photon in the laser beams. **2**

(iii) Assuming that all of the momentum of the photons is transferred to a rubidium atom, calculate the number of photons required to bring the atom to rest. **1**

(12)

[Turn over

Marks

11. A light source produces a beam of unpolarised light. The beam of light passes through a polarising filter called a polariser. The transmission axis of the polariser is shown in Figure 11A.

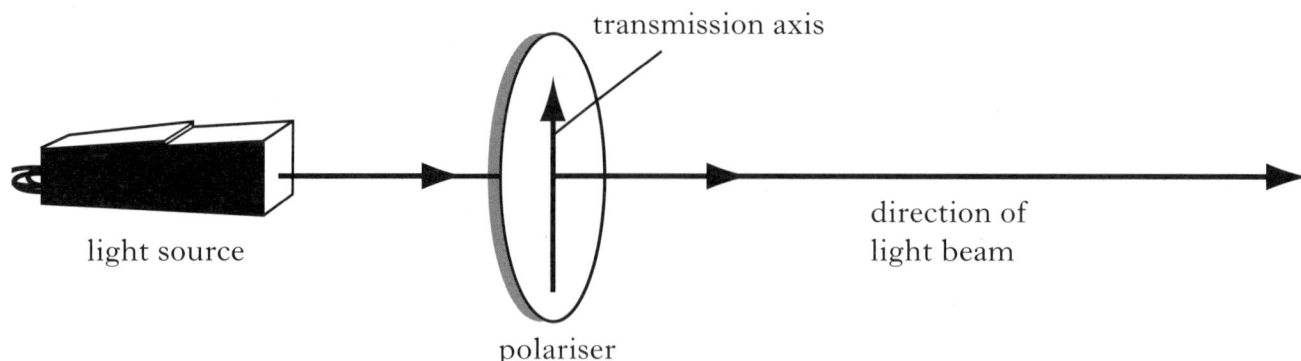

transmission axis

light source

direction of
light beam

polariser

Figure 11A

(a) Explain the difference between the unpolarised light entering the polariser and the plane polarised light leaving the polariser.

1

(b) The plane polarised light passes through a second polarising filter called an analyser.

The irradiance of the light passing through the analyser is measured by a light meter.

The transmission axis of the analyser can be rotated and its angle of rotation measured using a scale as shown in Figure 11B.

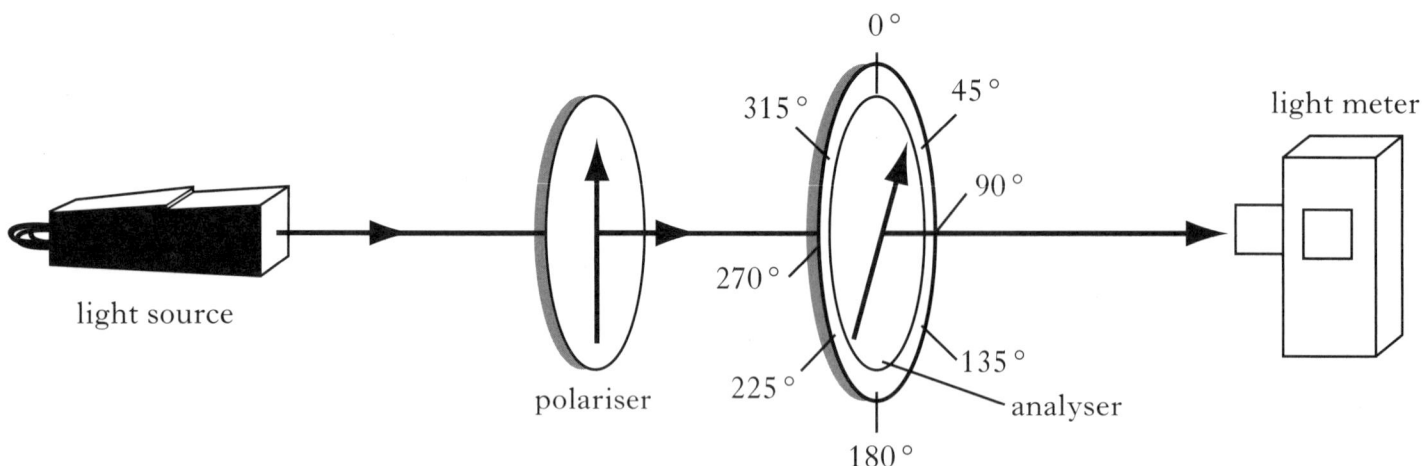

$0°$

$315°$ $45°$

light meter

$270°$ $90°$

$225°$ $135°$

light source

polariser   analyser

$180°$

Figure 11B

(i) The analyser is rotated.

State the **two** positions on the analyser scale that will produce a maximum reading of irradiance, I_0, on the light meter.

2

Marks

11. (b) (continued)

(ii) The relationship between the irradiance I detected by the light meter and the angle of rotation θ is given by

$$I = I_0 \cos^2\theta.$$

Explain how the equipment shown in Figure 11B could be used to establish this relationship.

Your answer should include:

- the measurements required;
- a description of how the relationship would be verified.

3

(6)

[END OF QUESTION PAPER]

[BLANK PAGE]

ADVANCED HIGHER

2011

[BLANK PAGE]

X069/701

NATIONAL	MONDAY, 23 MAY	PHYSICS
QUALIFICATIONS	1.00 PM – 3.30 PM	ADVANCED HIGHER
2011		

Reference may be made to the Physics Data Booklet.

Answer **all** questions.

Any necessary data may be found in the Data Sheet on page two.

Care should be taken to give an appropriate number of significant figures in the final answers to calculations.

Square-ruled paper (if used) should be placed inside the front cover of the answer book for return to the Scottish Qualifications Authority.

SQA

DATA SHEET

COMMON PHYSICAL QUANTITIES

Quantity	Symbol	Value	Quantity	Symbol	Value
Gravitational acceleration on Earth	g	$9{\cdot}8\ \mathrm{m\ s^{-2}}$	Mass of electron	m_e	$9{\cdot}11 \times 10^{-31}\ \mathrm{kg}$
Radius of Earth	R_E	$6{\cdot}4 \times 10^6\ \mathrm{m}$	Charge on electron	e	$-1{\cdot}60 \times 10^{-19}\ \mathrm{C}$
Mass of Earth	M_E	$6{\cdot}0 \times 10^{24}\ \mathrm{kg}$	Mass of neutron	m_n	$1{\cdot}675 \times 10^{-27}\ \mathrm{kg}$
Mass of Moon	M_M	$7{\cdot}3 \times 10^{22}\ \mathrm{kg}$	Mass of proton	m_p	$1{\cdot}673 \times 10^{-27}\ \mathrm{kg}$
Radius of Moon	R_M	$1{\cdot}7 \times 10^6\ \mathrm{m}$	Mass of alpha particle	m_α	$6{\cdot}645 \times 10^{-27}\ \mathrm{kg}$
Mean Radius of Moon Orbit		$3{\cdot}84 \times 10^8\ \mathrm{m}$	Charge on alpha particle		$3{\cdot}20 \times 10^{-19}\ \mathrm{C}$
Universal constant of gravitation	G	$6{\cdot}67 \times 10^{-11}\ \mathrm{m^3\ kg^{-1}\ s^{-2}}$	Planck's constant	h	$6{\cdot}63 \times 10^{-34}\ \mathrm{J\ s}$
Speed of light in vacuum	c	$3{\cdot}0 \times 10^8\ \mathrm{m\ s^{-1}}$	Permittivity of free space	ε_0	$8{\cdot}85 \times 10^{-12}\ \mathrm{F\ m^{-1}}$
Speed of sound in air	v	$3{\cdot}4 \times 10^2\ \mathrm{m\ s^{-1}}$	Permeability of free space	μ_0	$4\pi \times 10^{-7}\ \mathrm{H\ m^{-1}}$

REFRACTIVE INDICES

The refractive indices refer to sodium light of wavelength 589 nm and to substances at a temperature of 273 K.

Substance	Refractive index	Substance	Refractive index
Diamond	2·42	Glycerol	1·47
Glass	1·51	Water	1·33
Ice	1·31	Air	1·00
Perspex	1·49	Magnesium Fluoride	1·38

SPECTRAL LINES

Element	Wavelength/nm	Colour	Element	Wavelength/nm	Colour
Hydrogen	656	Red	Cadmium	644	Red
	486	Blue-green		509	Green
	434	Blue-violet		480	Blue
	410	Violet			
	397	Ultraviolet	*Lasers*		
	389	Ultraviolet	Element	Wavelength/nm	Colour
			Carbon dioxide	9550 ⎱ 10590 ⎰	Infrared
Sodium	589	Yellow	Helium-neon	633	Red

PROPERTIES OF SELECTED MATERIALS

Substance	Density/ $\mathrm{kg\ m^{-3}}$	Melting Point/ K	Boiling Point/ K	Specific Heat Capacity/ $\mathrm{J\ kg^{-1}\ K^{-1}}$	Specific Latent Heat of Fusion/ $\mathrm{J\ kg^{-1}}$	Specific Latent Heat of Vaporisation/ $\mathrm{J\ kg^{-1}}$
Aluminium	$2{\cdot}70 \times 10^3$	933	2623	$9{\cdot}02 \times 10^2$	$3{\cdot}95 \times 10^5$	. . . .
Copper	$8{\cdot}96 \times 10^3$	1357	2853	$3{\cdot}86 \times 10^2$	$2{\cdot}05 \times 10^5$	. . . .
Glass	$2{\cdot}60 \times 10^3$	1400	. . . .	$6{\cdot}70 \times 10^2$	. . . .	. . . .
Ice	$9{\cdot}20 \times 10^2$	273	. . . .	$2{\cdot}10 \times 10^3$	$3{\cdot}34 \times 10^5$	. . . .
Glycerol	$1{\cdot}26 \times 10^3$	291	563	$2{\cdot}43 \times 10^3$	$1{\cdot}81 \times 10^5$	$8{\cdot}30 \times 10^5$
Methanol	$7{\cdot}91 \times 10^2$	175	338	$2{\cdot}52 \times 10^3$	$9{\cdot}9 \times 10^4$	$1{\cdot}12 \times 10^6$
Sea Water	$1{\cdot}02 \times 10^3$	264	377	$3{\cdot}93 \times 10^3$	. . . .	. . . .
Water	$1{\cdot}00 \times 10^3$	273	373	$4{\cdot}19 \times 10^3$	$3{\cdot}34 \times 10^5$	$2{\cdot}26 \times 10^6$
Air	1·29	. . . .	. . . .	. . . .	. . . .	. . . .
Hydrogen	$9{\cdot}0 \times 10^{-2}$	14	20	$1{\cdot}43 \times 10^4$	. . . .	$4{\cdot}50 \times 10^5$
Nitrogen	1·25	63	77	$1{\cdot}04 \times 10^3$	. . . .	$2{\cdot}00 \times 10^5$
Oxygen	1·43	55	90	$9{\cdot}18 \times 10^2$	. . . .	$2{\cdot}40 \times 10^5$

The gas densities refer to a temperature of 273 K and a pressure of $1{\cdot}01 \times 10^5$ Pa.

Marks

1. In a process called "spallation", protons are accelerated to relativistic speeds and collide with mercury nuclei. Each collision releases neutrons from a mercury nucleus as shown in Figure 1.

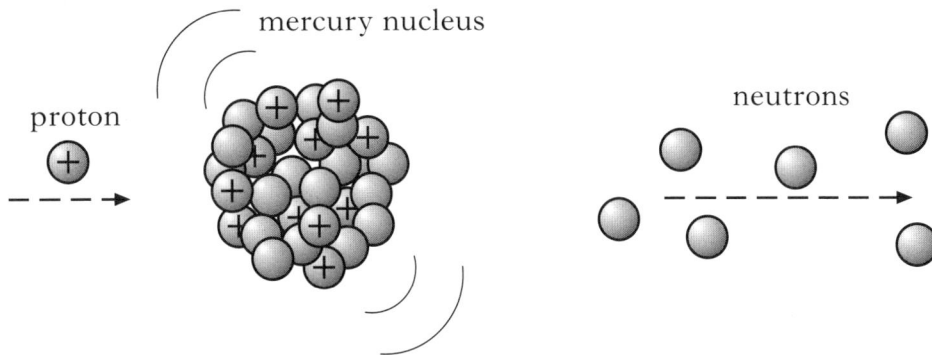

Figure 1

(a) (i) The energy of a proton is $2 \cdot 08 \times 10^{-10}$ J. Calculate the relativistic mass of this proton.   2

   (ii) Calculate the speed of this proton.   2

(b) A neutron produced in the spallation process is slowed to a non-relativistic speed, resulting in a kinetic energy of $3 \cdot 15 \times 10^{-21}$ J.

   (i) Show that the momentum of the neutron is $3 \cdot 25 \times 10^{-24}$ kg m s^{-1}.   2

   (ii) Calculate the de Broglie wavelength of this neutron.   2

(c) In a mercury nucleus, protons experience electrostatic repulsion, yet the nucleus remains stable.

   (i) Name the force responsible for this stability.   1

   (ii) Up to what distance is this force dominant?   1

   (iii) Name the fundamental particles that make up protons and neutrons.   1

                                            (11)

[Turn over

Marks

2. The front wheel of a racing bike can be considered to consist of 5 spokes and a rim, as shown in Figure 2A.

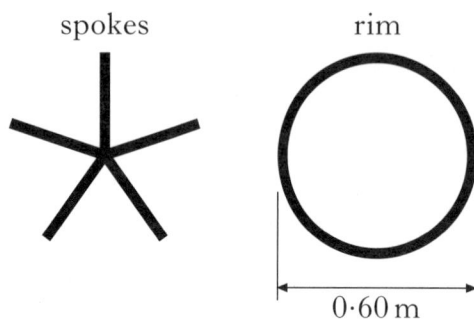

spokes   rim

0·60 m

Figure 2A

The mass of each spoke is 0·040 kg and the mass of the rim is 0·24 kg. The wheel has a diameter of 0·60 m.

(a) (i) Each spoke can be considered as a uniform rod. Calculate the moment of inertia of a spoke as the wheel rotates. **2**

(ii) Show that the total moment of inertia of the wheel is $2·8 \times 10^{-2}$ kg m^2. **2**

(b) The wheel is placed in a test rig and rotated as shown in Figure 2B.

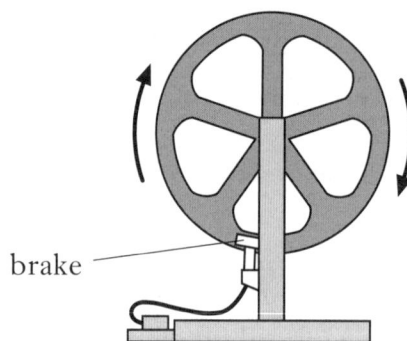

brake

Figure 2B

(i) The tangential velocity of the rim is 19·2 m s^{-1}. Calculate the angular velocity of the wheel. **2**

(ii) The brake is now applied to the rim of the wheel, bringing it uniformly to rest in 6·7 s.

(A) Calculate the angular acceleration of the wheel. **2**

(B) Calculate the torque acting on the wheel. **2**

(10)

Marks

3. An X-ray binary system consists of a star in a **circular** orbit around a black hole as shown in Figure 3A.

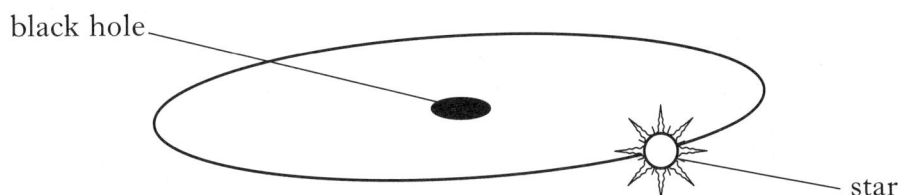

Figure 3A

The star has a mass of $2 \cdot 0 \times 10^{30}$ kg and takes 5·6 days to orbit the black hole. The orbital radius is $3 \cdot 6 \times 10^{10}$ m.

(a) Show that the angular velocity of the star is $1 \cdot 3 \times 10^{-5}$ rad s^{-1}.   **1**

(b) Calculate the mass of the black hole.   **3**

(c)   (i) Show that the potential energy of the star in its orbit is $-4 \cdot 4 \times 10^{41}$ J.   **1**

    (ii) Calculate the kinetic energy of the star.   **2**

    (iii) Calculate the total energy of the star due to its motion and position.   **1**

(d) The binary system orbits in the same plane as an earth-based telescope, as shown in Figure 3B.

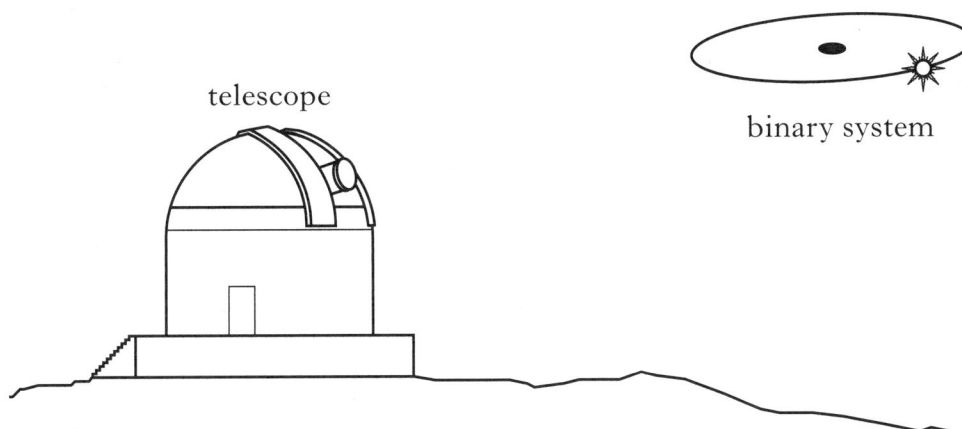

Figure 3B

Light from the star is analysed and found to contain the emission spectrum of hydrogen gas. The frequency of a particular line in the spectrum is monitored and a periodic variation in frequency is recorded.

Explain the periodic variation in the frequency.   **2**

(10)

[Turn over

Marks

4. A design for electrical power generation consists of a large buoy that drives a water column through a turbine as shown in Figure 4.

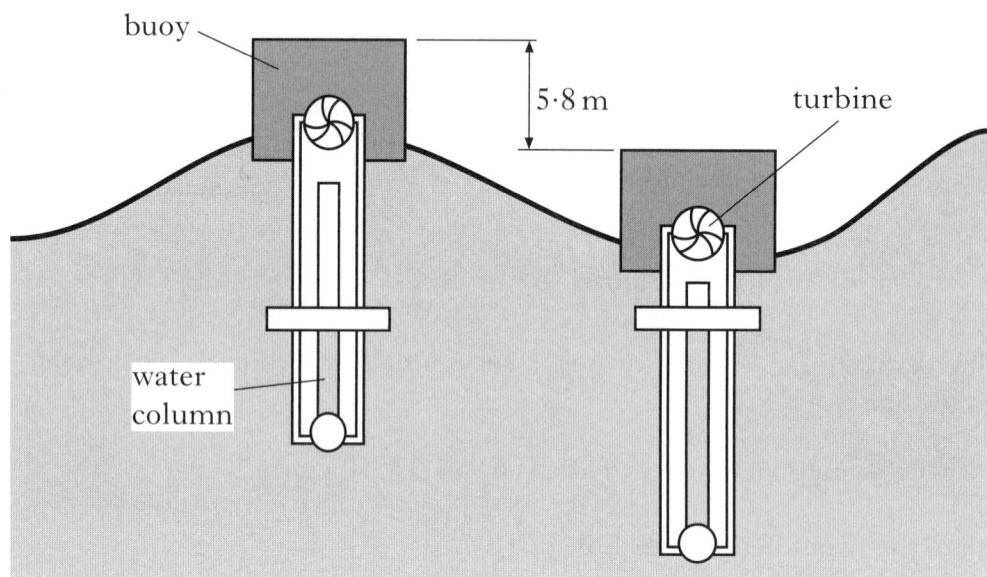

Figure 4

Energy is transferred from the wave motion to the turbines.

The mass of the buoy is 4.0×10^4 kg and its vertical displacement is 5.8 m. The motion of the buoy can be considered to be simple harmonic with a period of oscillation of 5.7 s.

(a) Write an equation that describes the vertical displacement y of the buoy. Numerical values are required.   **2**

(b) Calculate the maximum acceleration of the buoy.   **2**

(c) Where in the motion is the resultant force on the buoy greatest?   **1**

(d) Calculate the maximum kinetic energy of the buoy.   **2**

(e) The water column acts to damp the oscillatory motion of the buoy. How does this affect:

   (i) the period;   **1**

   (ii) the amplitude of the buoy's motion?   **1**

(9)

[Turn over for Question 5 on *Page eight*

Marks

5. A helium-filled metal foil balloon with a radius of $0.35\,m$ is charged by induction. The charge Q on the surface of the balloon is $+120\,\mu C$. The balloon is considered to be perfectly spherical.

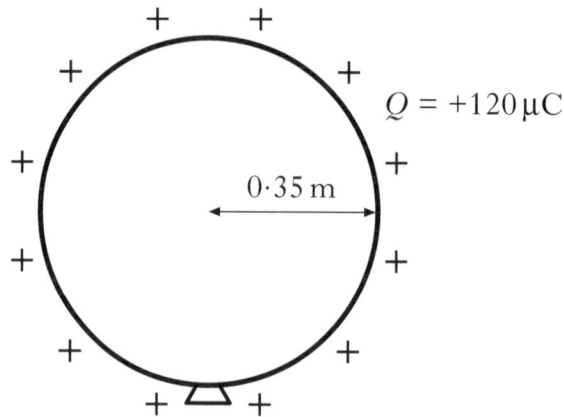

Figure 5A

(a) (i) Using diagrams, or otherwise, describe a procedure to charge the balloon by induction so that an evenly distributed positive charge is left on the balloon.   2

   (ii) Calculate the electric field strength at the surface of the balloon.   2

   (iii) Sketch a graph of the electric field strength against distance from the centre of the balloon to a point well beyond the balloon's surface. No numerical values are required.   1

(b) Two parallel charged plates are separated by a distance d. The potential difference between the plates is V.

   Lines representing the electric field between the plates are shown in Figure 5B.

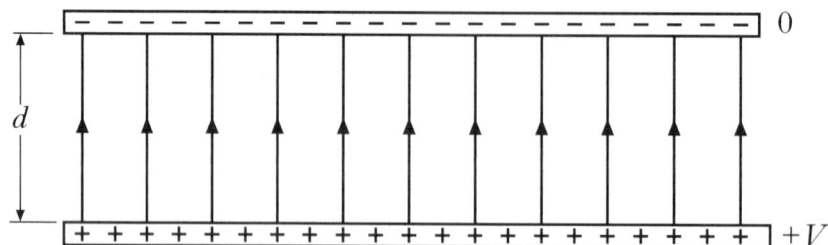

Figure 5B

   (i) By considering the work done in moving a point charge q between the plates, derive an expression for the electric field strength E between the plates in terms of V and d.   2

Marks

5. (b) (continued)

(ii) The base of a thundercloud is 489 m above an area of open flat ground as shown in Figure 5C.

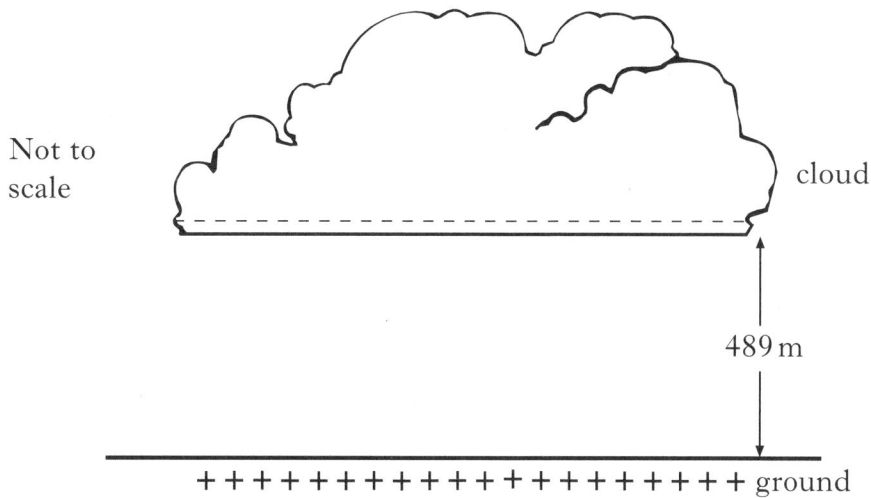

Figure 5C

The uniform electric field strength between the cloud and the ground is $7 \cdot 23 \times 10^4 \, \text{N C}^{-1}$.

Calculate the potential difference between the cloud and the ground.   **1**

(iii) During a lightning strike a charge of $5 \cdot 0 \, \text{C}$ passes between the cloud and the ground in a time of $348 \, \mu\text{s}$. The strike has negligible effect on the potential of the cloud. Calculate the average power of the lightning strike.   **2**

(c) An uncharged metal foil balloon is released and floats between the thundercloud and ground, as shown in Figure 5D.

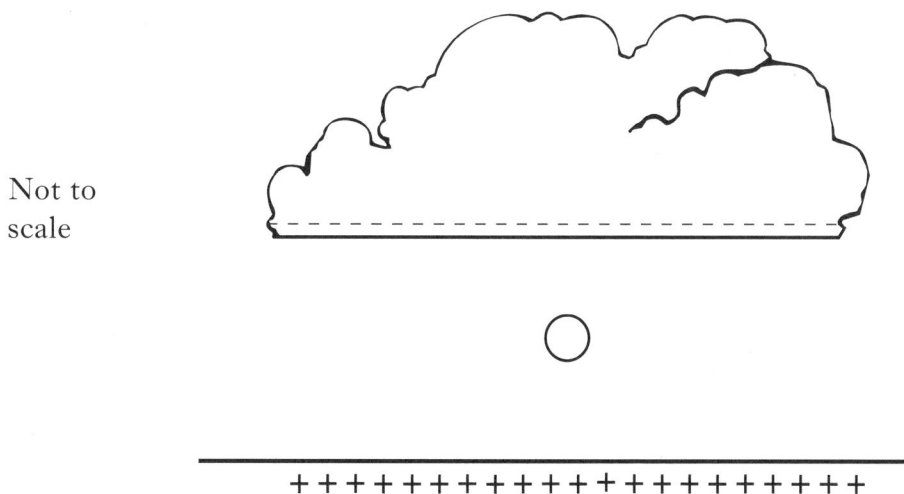

Figure 5D

Draw a diagram showing the charge distribution on the balloon and the resulting electric field around the balloon.   **2**

(12)

6. Modern trains have safety systems to ensure that they stop before the end of the line. One system being tested uses a relay operated by a reed switch. The reed switch closes momentarily as it passes over a permanent magnet laid on the track. An inductor in the relay activates the safety system as shown in Figure 6A.

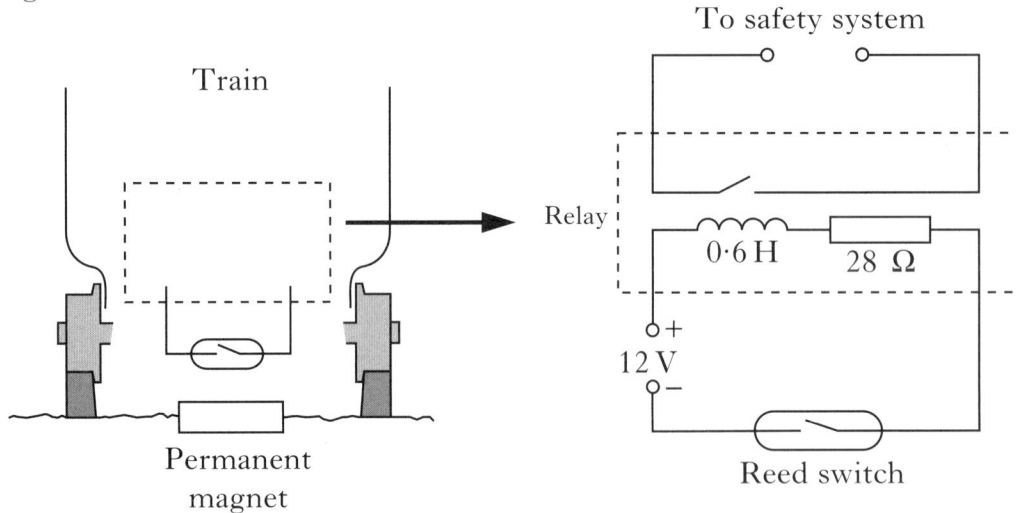

Figure 6A

(a) (i)   Explain why there is a short time delay between the reed switch closing and the relay activating.

2

   (ii) The inductor is connected to a 12·0 V d.c. supply. The inductor has an inductance of 0·6 H and the total resistance of the circuit is 28 Ω. Calculate the initial rate of change of current as the reed switch is closed.

2

   (iii) The inductance of the inductor on the train is 0·6 H. Define one henry.

1

   (iv) The reed switch opens as it moves away from the permanent magnet. Explain why a spark occurs across the contacts of the reed switch.

2

   (v)   A diode is placed across the inductor to prevent sparks across the reed switch as it opens as shown in Figure 6B. The diode must be chosen to carry the same current as the maximum current which occurs in the circuit when the reed switch closes. Calculate this current.

1

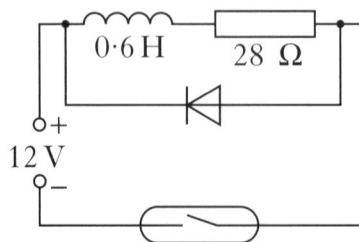

Figure 6B

Marks

6. (continued)

(b)   Another safety system prevents trains approaching a stop signal at excessive speed.   When a train is travelling too fast the brakes are applied automatically and the train is brought uniformly to rest. An inductor at the front of the train is used to determine the average speed as it travels between the electromagnets 1 and 2 as shown in Figure 6C.

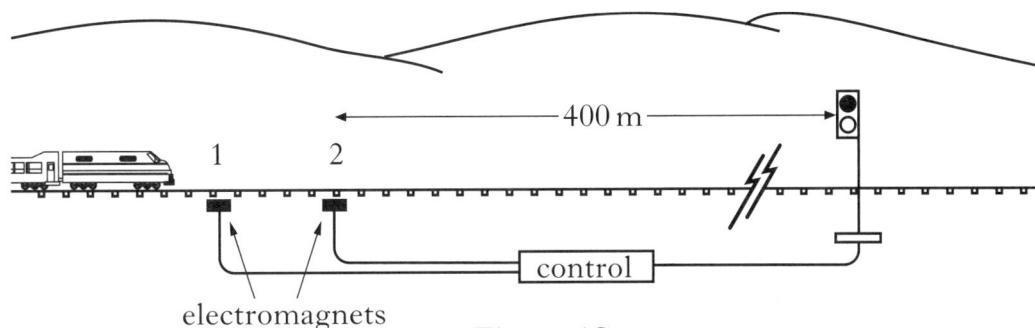

Figure 6C

The train travels between the electromagnets at a constant speed of $99\,km\,h^{-1}$. The brakes are applied automatically as the train passes the second electromagnet. The train is accelerated at $-1\cdot0\,m\,s^{-2}$. Show by calculation whether the train stops before the signal.

2

(c)   The train is stopped and a passenger hears a siren on another train approaching along a parallel track. The approaching train is travelling at a constant speed of $28\cdot0\,m\,s^{-1}$ and the siren produces a sound of frequency $294\,Hz$.

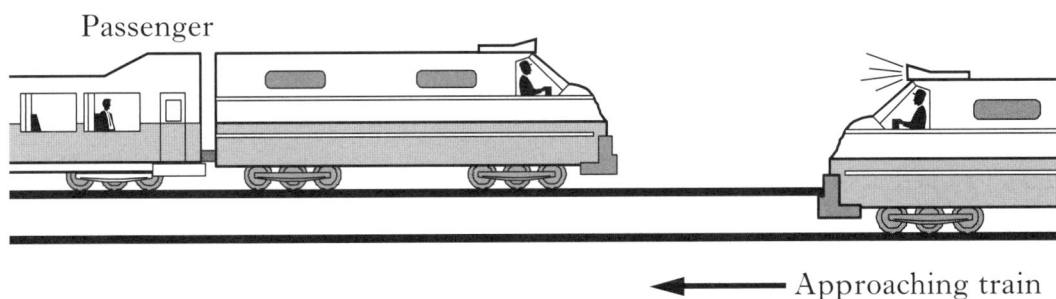

Figure 6D

(i)   Show that the frequency f of the sound heard by the passenger is given by

$$f = f_s\left(\frac{v}{v - v_s}\right)$$

where symbols have their usual meaning.

2

(ii)   Calculate the frequency of the sound heard by the passenger:

(A)   as the train approaches;

1

(B)   once the train has passed the passenger.

2

(15)

Marks

7. (a) Two very long straight wires X and Y are suspended parallel to each other at a distance r apart. The current in X is I_1 and the current in Y is I_2 as shown in Figure 7A.

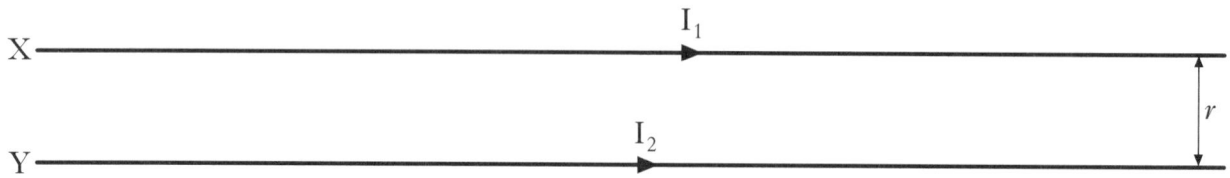

Figure 7A

(i) State the direction of the magnetic force acting on wire X. **Justify your answer**. 2

(ii) The wires are separated by a distance of 360 mm and each wire carries a current of 4·7A. Calculate the force per unit length which acts on each wire. 2

(b) A student investigating the force on a current carrying wire placed perpendicular to a uniform magnetic field obtains the following measurements and uncertainties.

Force (N)	0·0058 0·0061 0·0063 0·0057 0·0058 0·0062		
	Scale reading uncertainty	± 1 digit	
	Calibration uncertainty	± 0·00005 N	
Current (A)	Reading	1·98 A	
	Absolute uncertainty	± 0·02 A	
Length (m)	Reading	0·054 m	
	Absolute uncertainty	± 0·0005 m	

(i) From this data, calculate the magnetic induction, B. 3

(ii) Calculate the absolute uncertainty in the value of the force. 3

(iii) Calculate the overall absolute uncertainty in the value of the magnetic induction. 3

(13)

Marks

8. (*a*) Figure 8A shows a current carrying wire of length *l*, perpendicular to a magnetic field *B*. A single charge −*q* moves with constant velocity *v* in the wire. Using the relationship for the force on a current carrying conductor placed in a magnetic field, derive the relationship $F = qvB$ for the magnitude of the force acting on charge *q*.   **2**

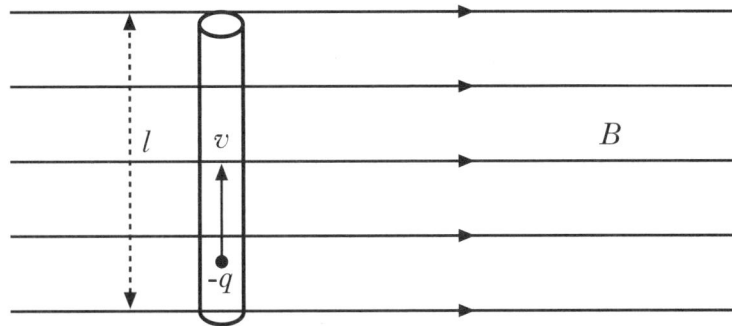

Figure 8A

(*b*) An electron with a speed of $2 \cdot 0 \times 10^6 \, \mathrm{m\,s^{-1}}$ enters a uniform magnetic field at an angle *θ*. The electron follows a helical path as shown in Figure 8B.

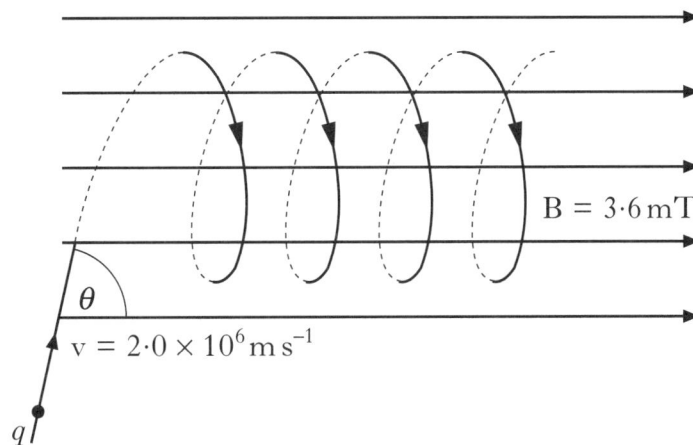

Figure 8B

The uniform magnetic induction is $3 \cdot 6 \, \mathrm{mT}$ and the radius of the helical path is $2 \cdot 8 \, \mathrm{mm}$. Calculate the value of angle *θ*.   **3**

(*c*) A second electron travelling at the same speed enters the field at a smaller angle *θ*.

Describe how the path of the second electron differs from the first.   **2**

   (7)

[Turn over

Marks

9. A laser-based quality control system to measure thread spacing in fabric samples is being evaluated. The 2-dimensional interference pattern is displayed on a screen shown in Figure 9A.

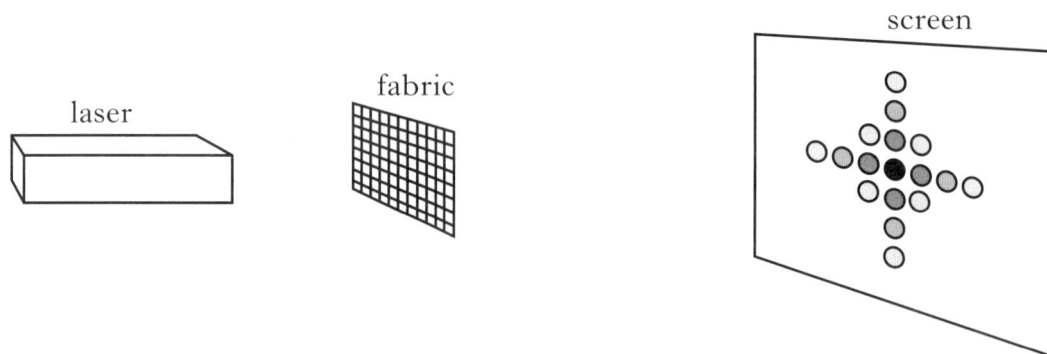

Figure 9A

(*a*) Explain how this 2-D interference pattern is produced.   **2**

(*b*) When a fine beam of laser light of wavelength 488 nm is used, the separation of the maxima in the horizontal direction is 8·00 mm. The distance from the fabric sample to the screen is 3·60 m.

Assume the spaces between the threads act like Young's slits.

Calculate the spacing between the threads in the sample.   **2**

Marks

9. (continued)

(c) The interference pattern from a standard fabric sample using a 488 nm laser is shown in Figure 9B.

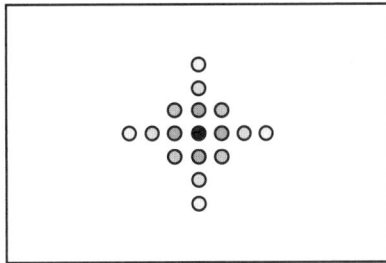

Figure 9B

(i) The 488 nm laser is replaced with a 667 nm laser. Which interference pattern from Figure 9C best represents the new interference pattern? Justify your answer.

2

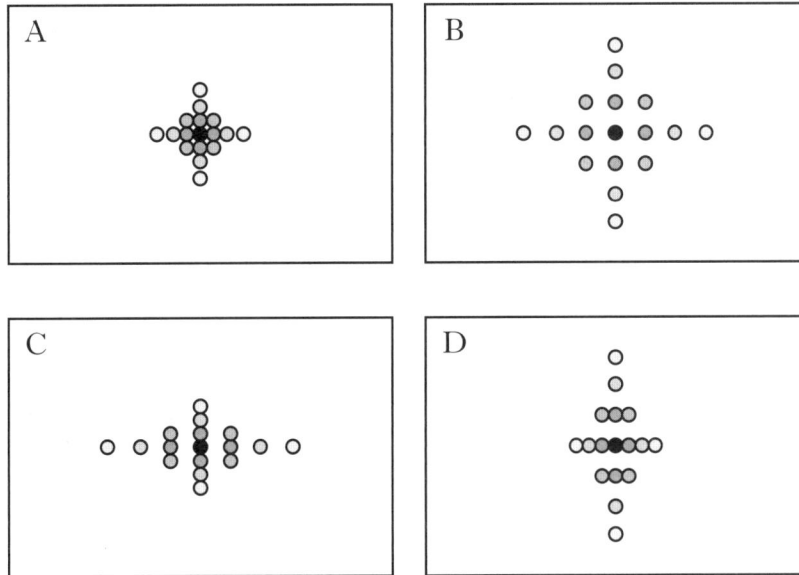

Figure 9C

(ii) The **original** 488 nm laser is restored and the fabric sample is stretched as shown in Figure 9D.

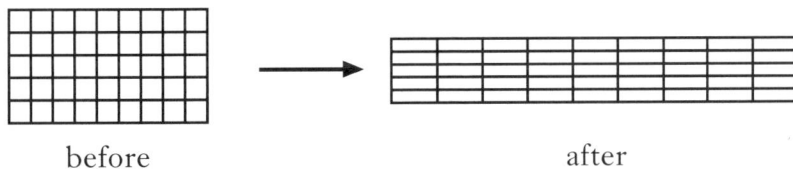

before   after

Figure 9D

Which pattern from Figure 9C best represents the new pattern? Justify your answer.

2

(8)

Marks

10. A stretched wire, supported near its ends, is made to vibrate by touching a tuning fork of unknown frequency to the supporting surface. One of the supports is moved until a stationary wave pattern appears as shown in Figure 10A.

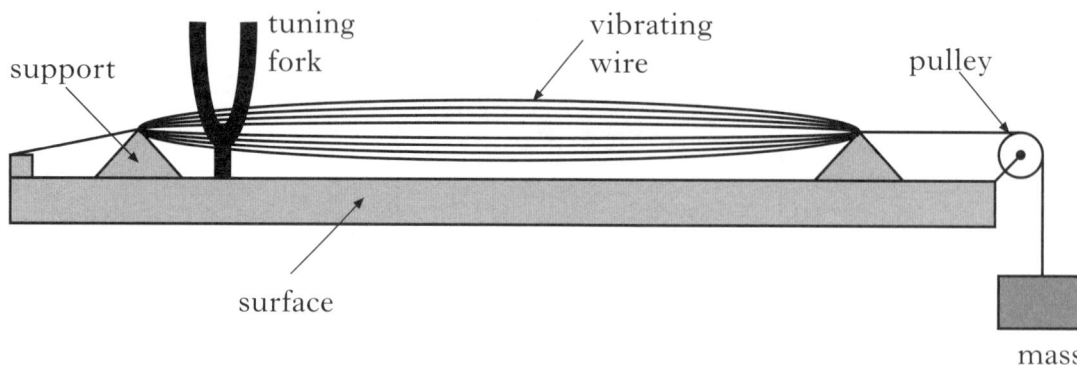

Figure 10A

(a) Explain how waves on this wire produce a stationary wave pattern.   **2**

(b) The formula for the frequency of the note from a stretched wire is given by:

$$f = \frac{1}{2l}\sqrt{\frac{T}{\mu}}$$

where l is the distance between the supports,
T is the stretching force,
μ is the mass per unit length of the wire.

The results of the experiment are given below:
mass per unit length of wire $= 1 \cdot 92 \times 10^{-4} \, kg \, m^{-1}$
distance between the supports $= 0 \cdot 780 \, m$
mass of load on wire $= 4 \cdot 02 \, kg$

(i) The table below gives information about the note produced by tuning forks of different frequency. Identify the note most likely to correspond to the tuning fork used in the experiment.   **2**

Note	A	B	C	D	E	F	G
Frequency (Hz)	220	245	262	294	330	349	392

Marks

10. (b) (continued)

(ii) A second tuning fork produces the pattern shown in Figure 10B. Suggest a frequency for this tuning fork.   **1**

(5)

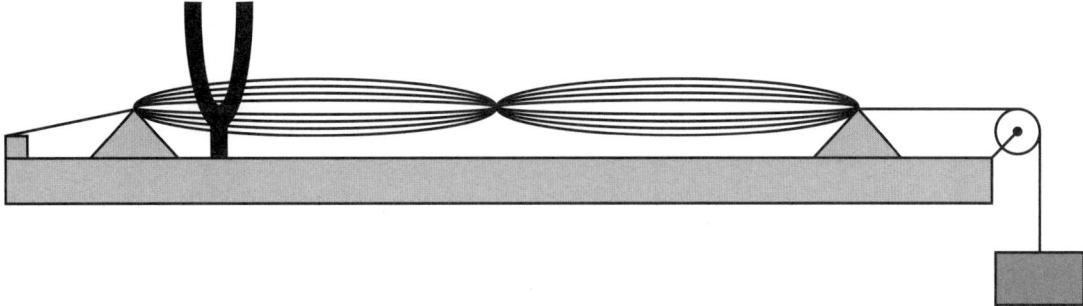

Figure 10B

[END OF QUESTION PAPER]

[BLANK PAGE]

ADVANCED HIGHER | ANSWER SECTION

SQA ADVANCED HIGHER PHYSICS 2007–2011

ADVANCED HIGHER PHYSICS 2007

1. (a) $\dfrac{ds}{dt} = v$

$\int ds = \int (u + at)\cdot dt$

$s = ut + \dfrac{1}{2}at^2 + c.$

at $t = 0$, $s = 0$, so $c = 0$.

$\therefore s = ut + \dfrac{1}{2}at^2$

(b) $4\cdot2 \times 10^{-10}$ J

2. (a) (i) $6\cdot8 \times 10^{-3}$ kgm^2

(ii) A. $\omega = \dfrac{\text{no. of revs}}{60} \times 2\pi$

$= \dfrac{45}{60} \times 2\pi$

$= 4\cdot7$ rad s^{-1}

B. $3\cdot1$ rad s^{-2}

(iii) $3\cdot6$ rad s^{-1}

(iv) Centripetal force is supplied by friction.
Force of friction is less than the required centripetal force.

(b) ω increases
r (of arms and legs) decreases
I decreases (inertia accepted)
Iω is constant

3. (a) (i) $2\cdot0 \times 10^{20}$ N
(ii) $v = 1\cdot0 \times 10^3$ m s^{-1}
(iii) work done in bringing unit mass from infinity (to a point in space)
(iv) $-7\cdot6 \times 10^{28}$ J
(v) $-4\cdot0 \times 10^{28}$ J

(b) (i) $E_P + E_K = 0$

$-\dfrac{GMm}{r} + \dfrac{1}{2}mv^2 = 0$

$\dfrac{1}{2}mv^2 = \dfrac{GMm}{r}$

$v = \sqrt{\dfrac{2GM}{r}}$

(ii) $2\cdot4 \times 10^3$ m s^{-1}

4. (a) force (or acceleration) is proportional to displacement and directed towards centre.

(b) $y = 0\cdot050 \cos 628t$

(c) $2\cdot0 \times 10^4$ ms^{-2}

(d) $9\cdot6 \times 10^3$ N

(e) 240 J

5. (a) (i) $V = \dfrac{Q}{4\pi\varepsilon_0 r}$

$= \dfrac{5\cdot1\times10^{-9}}{4\times3\cdot14\times8\cdot85\times10^{-12}\times0\cdot2}$

$= (230 \text{ V})$

(ii) 140 V

(b) bring rod close to sphere (or touching)
earth (touch) sphere.

(c) (i)

electric force

weight (or mg)

(ii) $1\cdot92 \times 10^{-18}$ C

(iii) 12

(d) not an integer multiple of e

6. (a) $B_{\perp} = B \sin 69$

$= 5 \times 10^{-5} \times \sin 69$

$= 4\cdot7 \times 10^{-5}$ T

(b) (i) $2\cdot1 \times 10^{-4}$ N

(ii) EAST

(c) 0 (N)

(d) (i) $1\cdot2 \times 10^{-2}$ m

(ii) circular

7. (a) (i) changing current causes changing magnetic field, producing back emf.

(ii) $0\cdot25$ A

(iii) $2\cdot4$ H

(iv) (A) smaller current
longer time

(b) B collapses (rapidly) **or** $\dfrac{dI}{dt}$ is large.

large (back) emf induced **or** emf > 80V

8. (a) (i) (A) $4\cdot8 \times 10^7$ m s^{-1}

(B) $8\cdot2 \times 10^7$ m s^{-1}

(ii) Perpendicular (component) - results in circular motion <u>or</u> central force.

Parallel (component) - constant velocity or no horizontal force or equivalent.

(b) (i) $\dfrac{mv^2}{r} = Bqv$

$\dfrac{m(v\sin\theta)^2}{r} = Bqv\sin\theta$

$r = \dfrac{mv\sin\theta}{Bq}$

$= \dfrac{9\cdot11 \times 10^{-31} \times 9\cdot5 \times 10^7 \times \sin60}{0\cdot22 \times 1\cdot6 \times 10^{-19}}$

$= 2\cdot1 \times 10^{-3}$m

(ii) $1\cdot6 \times 10^{-10}$ s

(iii) $7\cdot6 \times 10^{-3}$ m

(c) *Any two from:*

bigger radius

spirals in opposite sense (direction)

bigger pitch

9. (a) $y = 0 \cdot 05 \sin 2\pi \left(3t - \dfrac{x}{0 \cdot 02} \right)$

   (b) $y = 0 \cdot 05 \sin 2\pi \left(3t + \dfrac{x}{0 \cdot 02} \right)$

   (c) $0 \cdot 04$ m

10. (a) (i)   (A) π

         (B) π

      (ii)   (reflected rays) interfere destructively if optical path difference $= \dfrac{\lambda}{2}$

      (iii)   So more light is transmitted through the lens.

      (iv) $5 \cdot 80 \times 10^{-7}$m

   (b) fringe separation
      length of glass plates
      wavelength of sodium light
      $\Delta x = \dfrac{\lambda l}{2d}$

11. (a) $\lambda = 2 \times 0 \cdot 15$
      $v = f\lambda$
      $\quad = 250 \times 0 \cdot 3$
      $\quad (= 75 \text{ m s}^{-1})$

   (b) $v = (75 \pm 4) \text{ m s}^{-1}$

   (c) (i)   % uncertainty in λ will increase.

      (ii)   measure the distance over several nodes and take an average.

ADVANCED HIGHER PHYSICS 2008

1. (a) (i) 300 rad s^{-2}

      (ii) $0 \cdot 15$ Nm

      (iii) 380

   (b) (i) $0 \cdot 65$ N

      (ii)   (the force from) the glass/(end of) tube

   (c) increase
      r increases

2. (a) $(m)g = \dfrac{GM(m)}{R^2}$

      $R^2 = \dfrac{6 \cdot 67 \times 10^{-11} \times 6 \cdot 4 \times 10^{23}}{3 \cdot 7}$

      $\quad = 1 \cdot 15 \times 10^{13}$

      $R = \sqrt{1 \cdot 15 \times 10^{13}}$

      $\quad = 3 \cdot 4 \times 10^6$ m

   (b) (i) $\dfrac{GMm}{R^2} = m\omega^2 R$

             $\dfrac{GM}{R^2} = \omega^2 R$

             $\omega^2 = \dfrac{GM}{R^3}$

      (ii) $7 \cdot 1 \times 10^{-5}$ rads^{-1}

      (iii) $8 \cdot 9 \times 10^4$ s

   (c) $\dfrac{T^2}{R^3} = 3 \cdot 05 \times 10^{-7}$

      $\dfrac{T^2}{R^3} = 3 \cdot 02 \times 10^{-7}$

      $\dfrac{T^2}{R^3} = 3 \cdot 03 \times 10^{-7}$

      statement $\dfrac{T^2}{R^3} = $ constant

      or draw graph – gives straight line through origin of T^2 vs R^3

3. (a) (i) acceleration α – displacement

         or

         force α displacement and directed towards a fixed point

      (ii) $1 \cdot 5$ s

   (b) $8 \cdot 6 \times 10^{-2} \text{ m s}^{-1}$

   (c) $1 \cdot 8 \times 10^{-4}$ J

   (d) $0 \cdot 56$ m

   (e) **Amplitude** decreases

4. (a) Electrons exhibit $\begin{cases} \text{diffraction} \\ \text{interference} \end{cases}$

   (b) (i) $2 \cdot 1 \times 10^{-34}$kg m^2 s^{-1}

      (ii) $mrv = \dfrac{nh}{2\pi}$

$$mrv = \frac{2h}{2\pi}$$

$$2\pi r = \frac{2\,h}{mv}$$

$$= \frac{2h}{p}$$

$$= 2\lambda_B$$

(iii) $1\cdot1 \times 10^6 ms^{-1}$

5. (a) (i)

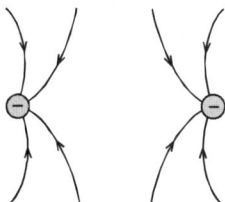

(ii) $V_X = -2\cdot4 \times 10^5$ V

(b) (i)

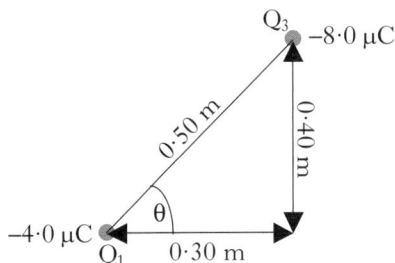

$$F = \frac{Qq}{4\pi\varepsilon_0 r^2}$$

$$F = \frac{-8\times10^{-6}\times-4\times10^{-6}}{4\times\pi\times8\cdot85\times10^{-12}\times0\cdot5^2}$$

$$= 1\cdot2\ N$$

(ii) $F = 2 \times 1\cdot2 \cos 37° = 1\cdot9$ N

Direction (000°)

6. (a) (i) gradient $= \dfrac{y_2 - y_1}{x_2 - x_1}$

$$\text{gradient} = \frac{3\cdot50 - 1\cdot50}{2\cdot25 - 1\cdot10}$$

$$= 1\cdot7 \times 10^{-3}\ NA^{-1}$$

(ii) $\pm0\cdot3 \times 10^{-3}\ NA^{-1}$

(Accept $0\cdot2 - 0\cdot3$)

(iii) $3\cdot3 \times 10^{-2} T$

(b) (i) Systematic uncertainty, calibration or zero error
(ii) *Any one from:*
  • take more readings (to increase n)
  • increase the range (to narrow parallelogram)
  • take multiple readings and average

7. (a) (i) A changing/increasing current in the inductor generates a back emf.

(ii) I = $0\cdot80$ A

(iii) E = $0\cdot26$ J

(iv) $13\ As^{-1}$

(b) (i) Maximum current unchanged

(ii) The time delay is decreased or the time to reach maximum current is reduced, because the inductance is decreased (by removing the iron core) (**or** back emf is reduced)

(iii) I is the same but L is smaller.

(c) I

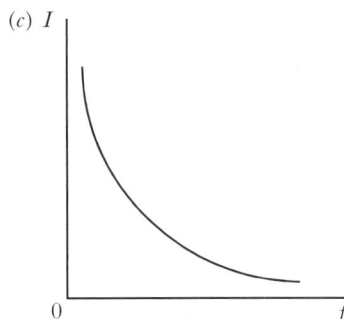

8. (a) (i) $F_{(E)}\dfrac{Q_1 Q_2}{4\pi\varepsilon_0 r^2} = 4\cdot8\times10^{-19}(N)$

$$F_{(G)} = \frac{Gm_1 m_2}{r^2}$$

$$F_G = 3\cdot9 \times 10^{-55}\ (N)$$

$F_G(10^{-55}N) << F_E(10^{-19}N)$

(ii) Strong force only acts at a range of approx. 10^{-14} m.

or

The distance between these 2 protons is too large.

(b) $E_K = E_p$

$$\therefore \frac{1}{2}mv^2 = \frac{Qq}{4\pi\varepsilon_0 r_c}$$

$$v^2 = \frac{2Qq}{4\pi\varepsilon_0 m r_c}$$

$$v = \sqrt{\frac{Qq}{2\pi\varepsilon_0 m r_c}}$$

(c) (i) $Q = 1\cdot2 \times 10^{-17}$ C

(ii) 75

(iii) It is a Rhenium nucleus (atomic number 75)

9. (a) (i) frequency increased (with moving observer)
Driver passing through more (than 1250) wavefronts in <u>1 second</u>

(ii) 1340 Hz

(b) moving away
$\begin{Bmatrix} \text{wavelength increased} \\ \textbf{or}\ \text{frequency decreased} \end{Bmatrix}$

10. (a) (i) Polarised light: (The electric field vector of) the wave **oscillates** or **vibrates** in one **plane.**

(ii) $\mu = \dfrac{\sin i}{\sin r}$

$$\mu = \frac{\sin 34\cdot0}{\sin 48\cdot0} = 0\cdot752$$

or

$$\mu = \frac{1}{n}$$

$$\mu = \frac{1}{1\cdot33} = 0\cdot752$$

    (iii)   36·9°

(b) 0·2°

(c) Intensity changes

    In a cyclic fashion

11. (a) Division of wavefront

(b) $3·1 \times 10^{-4}$m

(c) (i)   increased Δx
      Smaller % uncertainty in d or Δx

   (ii)   Fainter fringes
      or broader fringes
      or not all fringes seen, screen not big enough

ADVANCED HIGHER PHYSICS 2009

1. (a) (i) $v = \dfrac{ds}{dt}$

      $= 6·2t + 4·1$

  (ii) $72 = 6·2t + 4·1$

      $t = \dfrac{72 - 4·1}{6·2}$

      $= 11\text{s}$

 (iii) $a = \dfrac{dv}{dt}$

      $= 6·2\,m\,s^{-2}$

(b) (i)   Escape velocity greater than c or
      3×10^8 m s^{-1} **or** no light can escape

  (ii)   The escape velocity is the minimum velocity an object must have which would allow it to escape the gravitational field.

 (iii) $E_P + E_K = 0$

      $\dfrac{-GMm}{r} + \frac{1}{2}\,mv^2 = 0$

      $\frac{1}{2}\,mv^2 = \dfrac{GMm}{r}$

      $v = \sqrt{\dfrac{2GM}{r}}$

 (iv) $v_e = c = \sqrt{\dfrac{2GM}{r}}$

      $3·0 \times 10^8 = \sqrt{\dfrac{2 \times 6·67 \times 10^{-11} \times 4·58 \times 10^{30}}{r}}$

      $r = 6·8 \times 10^3$ m

  (v) $\rho = \dfrac{M}{V} \quad V = \dfrac{4}{3}\pi r^3$

      $\rho = \dfrac{4·58 \times 10^{30}}{\frac{4}{3}\pi \times \left(6·8 \times 10^3\right)^3}$

      $= 3·5 \times 10^{18}$ kg m^{-3}

2. (a) (i) $\omega = (48 \times 5·8) - 12 = 266$ rpm

      or from the graph taking $\omega = 265$ rpm

      $\omega = \dfrac{266 \times 2\pi}{60} = 28$ rad s^{-1}

  (ii) $\omega = (48 \times 1·6) - 12 = 65$ (rpm)

      $\omega = \dfrac{65 \times 2\pi}{60} = 6·8$ (rad s^{-1})

      $\alpha = \dfrac{\omega - \omega_0}{t}$

      $\alpha = \dfrac{6·8 - 28}{8}$

      $\alpha = \dfrac{6·8 - 28}{8} = -2·7$ rad s^{-2}

(b) (i) $I = \frac{1}{3}\,ml^2$

      $= \frac{1}{3} \times 11 \times 10^{-3} \times (76 \times 10^{-3})^2$

      $= 2·1 \times 10^{-5}$ kg m^2

  (ii) $I_{total} = 3I + I_{cylinder}$

      $= (3 \times 2·1 \times 10^{-5}) + 1·1 \times 10^{-6}$

      $= 6·4 \times 10^{-5}$ kg m^2

(c) $T = I\alpha$

$\quad = 6{\cdot}4 \times 10^{-5} \times 2{\cdot}7$

$\quad = 1{\cdot}7 \times 10^{-4}\,\text{N m}$

(d) Moment of inertia would increase and then one from the following: greater time to stop/α would decrease/speed of rotation would be less

3. (a) (i) $\omega = 2\pi f$

$\quad = 2 \times \pi \times 33 = 210$

$y = 2{\cdot}1 \times 10^{-3} \sin 210t$ **or**

$y = 2{\cdot}1 \times 10^{-3} \cos 210t$

(ii) $v_{max} = \pm\omega A$

$v_{max} = \pm 210 \times 2{\cdot}1 \times 10^{-3}$

$v_{max} = \pm 0{\cdot}44\,\text{m s}^{-1}$

(b) $\omega = 77\,\text{rad s}^{-1}$

$A\omega = 9{\cdot}2 \times 10^{-2}$

$\therefore A = \dfrac{9{\cdot}2 \times 10^{-2}}{77} = 1{\cdot}2 \times 10^{-3}\,\text{m}$

(c) (i) $E_{k\,max} = \dfrac{1}{2}\,m\omega^2 A^2$

From equation $\omega = 77\,\text{rad s}^{-1}$,

$A = 1{\cdot}2 \times 10^{-3}\,\text{m}$

$E_{k\,max} = \dfrac{1}{2} \times 2{\cdot}5 \times 10^{-6} \times 77^2 \times (1{\cdot}2 \times 10^{-3})^2$

$\quad = 1{\cdot}1 \times 10^{-8}\,\text{J}$

(ii)

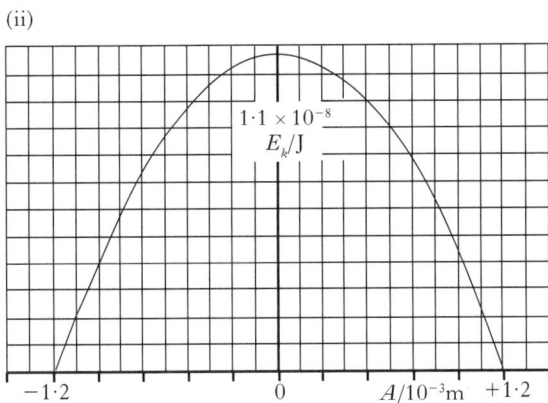

4. (a) (i)

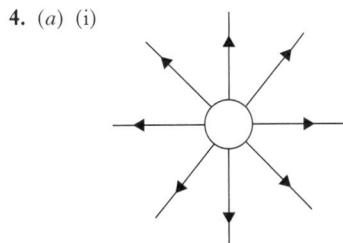

(ii) E field/force from A and from B are in same direction/add up/don't cancel

or

Diagram showing field lines running between A and B – must have arrow

Direction of arrow $+$ to $-$

(iii) $E_A = \dfrac{q}{4\pi\varepsilon_0 r^2}$

$\quad = \dfrac{4 \times 10^{-6}}{4 \times 3{\cdot}14 \times 8{\cdot}85 \times 10^{-12} \times 0{\cdot}34^2}$

$\quad = 3{\cdot}1 \times 10^5\,(\text{N C}^{-1})$

$E_B = \dfrac{q}{4\pi\varepsilon_0 r^2}$

$\quad = \dfrac{-2 \times 10^{-6}}{4 \times 3{\cdot}14 \times 8{\cdot}85 \times 10^{-12} \times 0{\cdot}24^2}$

$\quad = -3{\cdot}1 \times 10^5\,(\text{N C}^{-1})$

$\Rightarrow E_{total} = 0\,(\text{N C}^{-1})$

(b) The strong force

Balances/greater than the repulsive/electrostatic force **or** this force acts over a short range.

(c) (i) charge $2 \times {}^2/_3 + 1 \times -{}^1/_3 = e$

or

baryon $2 \times {}^1/_3 + 1 \times {}^1/_3 = 1$

(ii) Down quark & anti up quark

5. (a) $F = qvB$

$5 \times 10^{-11} = 3{\cdot}2 \times 10^{-19} \times v \times 6{\cdot}8$

$v = 2{\cdot}3 \times 10^7\,\text{m s}^{-1}$

(b) $v = \dfrac{E}{B}$

$2{\cdot}3 \times 10^7 = \dfrac{E}{6{\cdot}8}$

$E = 1{\cdot}6 \times 10^8\,\text{V m}^{-1}$

or

$E = \dfrac{F}{q}$

$E = \dfrac{5{\cdot}0 \times 10^{-11}}{3{\cdot}2 \times 10^{-19}}$

$E = 1{\cdot}6 \times 10^8\,\text{N C}^{-1}$

(c) $F = \dfrac{mv^2}{r}$

$5{\cdot}0 \times 10^{-11} = \dfrac{6{\cdot}645 \times 10^{-27} \times (2{\cdot}3 \times 10^7)^2}{r}$

$r = 0{\cdot}070\,(\text{m})$

alpha particle hits at position B/$0{\cdot}14$ m

(d) Electron will be deflected in the opposite direction due to opposite charge.

Radius of semicircle smaller due to (much) smaller mass

or greater $\dfrac{q}{m}$.

6. (a) (i)

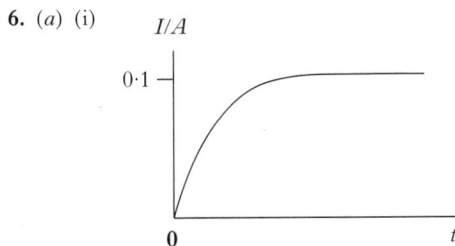

(ii) Max voltage is 3 V

or Back emf too small

(iii) Magnetic field <u>collapse/falls quickly</u>

Large (back) emf (110 V produced)

(iv) $E = Pt$

$\qquad = 1 \cdot 2 \times 10^{-3} \times 0 \cdot 25$

$\qquad = 3 \times 10^{-4} \text{ (J)}$

$E = \frac{1}{2} LI^2$

$3 \times 10^{-4} = \frac{1}{2} \times L \times 0 \cdot 1^2$

$L = 0 \cdot 060 \text{ H}$

(b) (i) Voltmeter is to monitor voltage so voltage across inductor remains <u>constant</u>

(ii) Draw a graph of I against $1/f$

or

Check $I \times f$ remains constant for all values

(iii) $I \times f = $ constant

or

$I \propto 1/f$

(c) LS1 will produce low frequency sounds (woofer)
or LS2 will produce high frequency sounds (tweeter).
At high frequency capacitive reactance is low.
At low frequency inductive reactance is low.

7. (a) (i) $L = \frac{nh}{2\pi}$

$\qquad = \frac{1 \times 6 \cdot 63 \times 10^{-34}}{2\pi}$

$L = 1 \cdot 06 \times 10^{-34} \text{ kg m}^2 \text{ s}^{-1}$ or kg m^2 rad s^{-1}
or J s

(ii) $mv = \frac{nh}{2\pi r}$

$mv = \frac{1 \times 6 \cdot 63 \times 10^{-34}}{5 \cdot 3 \times 10^{-11} \times 2\pi}$

$mv = 2 \cdot 0 \times 10^{-24} \text{ kg m s}^{-1}$

OR

$mv = \frac{L}{r}$

$mv = \frac{1 \cdot 06 \times 10^{-34}}{5 \cdot 3 \times 10^{-11}}$

$mv = 2 \cdot 0 \times 10^{-24} \text{ kg m s}^{-1}$

(iii) $\lambda = \frac{h}{p}$

$\qquad = \frac{6 \cdot 63 \times 10^{-34}}{2 \cdot 0 \times 10^{-24}}$

$\qquad = 3 \cdot 3 \times 10^{-10} \text{ m}$

(b) (i) The electrons would spiral inwards towards the nucleus.

or Orbit decays / decreases

(ii) Quantum mechanics.

8. (a) $B = \frac{\mu_0 I}{2\pi r}$

$1 \cdot 7 \times 10^{-7} = \frac{4 \times \pi \times 10^{-7} \times I}{2 \times \pi \times 0 \cdot 25}$

$I = 0 \cdot 21 \text{ A}$

(b) One tesla is the magnetic induction of a magnetic field in which a conductor of length **one metre**, carrying a current of **one ampere** perpendicular to the field is acted on by a force of **one newton**.

(c) $F = BIl$

$F/l = 1 \cdot 7 \times 10^{-7} \times 2$

$\qquad = 3 \cdot 4 \times 10^{-7} \text{ N (m}^{-1})$

or

$F/l = \frac{\mu_0 I_1 I_2}{2\pi r}$

$\qquad = \frac{4 \times \pi \times 10^{-7} \times 0 \cdot 21 \times 2}{2 \times \pi \times 0 \cdot 25}$

$\qquad = 3 \cdot 4 \times 10^{-7} \text{ N (m}^{-1})$

9. (a) (i) Division of amplitude is when some of the light **reflects** from the top of the air wedge and some is **transmitted/refracted** into the air.

or

Some of the light is **reflected** from a surface of a new material/medium and some of the light is **transmitted/refracted** into the new material/medium.

(ii) $10 \Delta x = 6 \cdot 0 \times 10^{-4}$

$\Delta x = 6 \cdot 0 \times 10^{-5} \text{ (m)}$

$\Delta x = \frac{\lambda l}{2d}$

$6 \cdot 0 \times 10^{-5} = \frac{580 \times 10^{-9} \times 4 \cdot 0 \times 10^{-2}}{2d}$

$d = 1 \cdot 9 \times 10^{-4} \text{ m}$

(iii) $\% \Delta x = \frac{0 \cdot 5 \times 100}{6 \cdot 0} \quad 8 \cdot 3 \text{ (\%)}$

$\% \Delta \lambda = \frac{10 \times 100}{580} = 1 \cdot 7 \text{ (\%)}$

$\% \Delta l = \frac{0 \cdot 1 \times 100}{4 \cdot 0} = 2 \cdot 5 \text{ (\%)}$

$\% \Delta d = 8 \cdot 3 \text{ (\%)}$

(b) (i) Light is reflected from both surfaces of the soap film. The two reflected waves meet out of phase by π or $\lambda/2$.

(ii) $opd. = 2 \times thickness \times n$

$\qquad = 2 \times 4 \cdot 00 \times 10^{-6} \times 1 \cdot 45$

$\qquad = 1 \cdot 16 \times 10^{-5} \text{ m}$

(iii) The next point giving destructive interference must have:-
an optical path difference of one λ more than that at X.
$New\ opd = 1 \cdot 16 \times 10^{-5} + 580 \times 10^{-9}$

$\qquad = 1 \cdot 22 \times 10^{-5} \text{ m}$

10. (a) A stationary wave is caused by **interference** effects between the incident and **reflected** sound.

(b) The antinodes of the pattern are areas of **maximum** displacement/amplitude/disturbance

The nodes of the pattern are areas of **minimum/zero** displacement/amplitude/ disturbance

(c) The beads accumulate at the nodes - the vibrations at the antinodes pushes them to side.

(d) $\lambda = 2 \times 85 \times 10^{-3}$

$\qquad = 170 \times 10^{-3} \text{ (m)}$

$v = f\lambda$

$\qquad = 1950 \times 170 \times 10^{-3}$

$v = 330 \text{ m s}^{-1}$

ADVANCED HIGHER PHYSICS 2010

1. (a) $v = \dfrac{s}{t}$ **or** $v = \dfrac{s}{t}$

$v = \dfrac{2\pi r}{T}$ $v = \dfrac{r\theta}{t}$

$\omega = \dfrac{2\pi}{T}$ $\omega = \dfrac{\theta}{t}$

$v = r\omega$ $v = r\omega$

(b) (i) $\omega = \dfrac{\theta}{t}$

$= \dfrac{3\cdot1}{4\cdot5} = 0\cdot69$

$v = r\omega$
$= 0\cdot69 \times 0\cdot148$
$= 0\cdot10\ \text{m s}^{-1}$

(ii) $\%\Delta\theta = \dfrac{0\cdot1}{3\cdot1} \times 100 = 3\cdot2\%$

$\%\Delta t = \dfrac{0\cdot1}{4\cdot5} \times 100 = 2\cdot2\%$

$\%\Delta r = \dfrac{0\cdot001}{0\cdot148} \times 100 = 0\cdot68\%$

$\%\Delta\omega = \sqrt{(\%\Delta\theta^2 + \%\Delta t^2 + \%\Delta r^2)}$
$= \sqrt{(3\cdot2^2 + 2\cdot2^2 + 0\cdot68^2)}$
$= 3\cdot9\ (\%)$

(iii) (A) velocity changing
or
changing direction
(B) towards centre (of turntable)

2. (a) $F = m r\omega^2$
$= 0\cdot2 \times 0\cdot35 \times 6\cdot0^2$
$= 2\cdot5\ \text{N}$

(b) $\tan\theta = \dfrac{m\omega^2 r}{mg}$ or $= \dfrac{\omega^2 r}{g}$ or $= \dfrac{v^2}{rg}$

$= \dfrac{2\cdot5}{0\cdot2 \times 9\cdot8}$ **or** $\dfrac{6\cdot0^2 \times 0\cdot35}{9\cdot8}$ **or** $\dfrac{2\cdot1^2}{0\cdot35 \times 9\cdot8}$

$\theta = 52°$

(c) θ decreases
Centripetal force or $r\omega^2$ or v^2 / r decreases

3. (a) $\alpha = \dfrac{\omega - \omega_0}{t}$

$= \dfrac{35 - 0}{0\cdot55}$

$= 64\ \text{rad s}^{-2}$

(b) θ = area under graph **or** $a_t = r\alpha$
$= \frac{1}{2} \times b \times h$ $= 0\cdot14 \times 64$
$= \frac{1}{2} \times 0\cdot55 \times 35$ $= 8\cdot96\ \text{ms}^{-2}$
$= 9\cdot6\ (\text{rad})$ $s = ut + \frac{1}{2}at^2$
$s = r \times \theta$
$= 0\cdot14 \times 9\cdot6$ $= 0 + \frac{1}{2} \times 8\cdot96 \times 0\cdot55^2$
$= (1\cdot3\ \text{m})$ $= 1\cdot36\ \text{m (accept)}$
rounding allowance

(c) $mgh = 2\cdot5 \times 9\cdot8 \times 1\cdot3$
$v = \omega \times r$
$= 35 \times 0\cdot14$
$= 4\cdot9\ (\text{ms}^{-1})$

$mgh = \frac{1}{2}mv^2 + \frac{1}{2}I\omega^2$

$2\cdot5 \times 9\cdot8 \times 1\cdot3 = \frac{1}{2} \times 2\cdot5 \times 4\cdot9^2 + \frac{1}{2} \times I \times 35^2$

$31\cdot85 = 30\cdot01 + 612\cdot5 \times I$

$I = \dfrac{1\cdot84}{612\cdot5}$

$= 3\cdot0 \times 10^{-3}\ \text{kg m}^2$

4. (a) **Total angular** momentum before (an event) = **total angular** momentum after (an event)

in the absence of external torques

(b) (i) $L = I \times \omega$
$= 4\cdot1 \times 2\cdot7$
$= 11\ \text{kg m}^2\ \text{s}^{-1}$ or $\text{kg m}^2\ \text{rad s}^{-1}$

(ii) $I_m = m r^2 = (2\cdot5 \times 0\cdot60^2) = \mathbf{0\cdot90}\ \text{kg m}^2$
$I_T = (4\cdot1 + 0\cdot90)$ $= \mathbf{5\cdot0}\ \text{kg m}^2$
$I_0 \times \omega_0 = I_T \times \omega_T$
$11 = 5\cdot0 \times \omega$
$\omega = 2\cdot2\ \text{rad s}^{-1}$

(iii) ω increased
r reduced / I reduced
since L or $I\omega$ constant

(c) $\alpha = \dfrac{\omega - \omega_0}{t}$

$= \dfrac{0 - 1\cdot5}{0\cdot75}$

$= -2\cdot0\ \text{rad s}^{-2}$

$\tau = I \times \alpha$
$= 4\cdot5 \times -2$
$= -9\ \text{N m}$

5. (a) $a = -\omega^2 y$
$-35 = -\omega^2 \times 0\cdot012$ **or** $35 = -\omega^2 \times (-0\cdot012)$
$\omega = 54\ \text{rad s}^{-1}$

(b) $y = 0\cdot012\ \sin$ or $\cos 54t$

(c) $v = \dfrac{dy}{dt}$ **or** $\dfrac{d(0\cdot012\ \sin 54t)}{dt}$

$v = (+)\ 0\cdot65\ \cos 54t$
or
$v = -0\cdot65\ \sin 54t$

(d) $E_k = \frac{1}{2}m\omega^2 A^2$
$= \frac{1}{2} \times 1\cdot4 \times 54^2 \times 0\cdot012^2$
$= 0\cdot29\ \text{J}$

6. (a) $E = \dfrac{Q}{4\pi\epsilon_o r^2}$

$E = \dfrac{-1\cdot92 \times 10^{-12}}{4 \times \pi \times 8\cdot85 \times 10^{-12} \times (1\cdot00 \times 10^{-3})^2}$

$E = -1\cdot73 \times 10^4\ \text{NC}^{-1}$

(b) (i) Student A
(ii) E-field is zero inside a hollow conductor
E-field has inverse square dependence
outside the conductor

(c) (i) $F = \dfrac{Q_1 Q_2}{4\pi\epsilon_o r^2}$

$F = \dfrac{(-)2\cdot97 \times 10^{-8} \times (-)1\cdot92 \times 10^{-12}}{4 \times \pi \times 8\cdot85 \times 10^{-12} \times (4\cdot12 \times 10^{-2})^2}$

$F = 3\cdot02 \times 10^{-7}\ \text{N}$

(ii)

$\sin\theta = \dfrac{30\cdot0}{41\cdot2}$

$\theta = 46\cdot7°$

$F = F \times \sin\theta$
$F = 3\cdot02 \times 10^{-7} \times \sin 46\cdot7°$
$F = 2\cdot20 \times 10^{-7}$ N

(iii) Resultant $F = 4 \times F$
$= 8\cdot80 \times 10^{-7}$ N

(iv) Resultant $F = m \times g$
$8\cdot80 \times 10^{-7} = m \times 9\cdot8$
$m = 9\cdot0 \times 10^{-8}$ kg

7. (a) Magnetic field is out of the page

(b) $q \times v \times B = m \times \dfrac{v^2}{r}$

$r = \dfrac{m \times v}{q \times B}$

$= \dfrac{1\cdot673 \times 10^{-27} \times 6\cdot0 \times 10^6}{1\cdot6 \times 10^{-19} \times 0\cdot75}$

$r = 8\cdot4 \times 10^{-2}$m

(c) $q \times E = q \times v \times B$
$E = v \times B$
$= 6\cdot0 \times 10^6 \times 0\cdot75$
$E = 4\cdot5 \times 10^6$ Vm^{-1}

(d) (i) $\frac{1}{2} \times m \times v^2 = \dfrac{qQ}{4\pi\epsilon_0 r}$

$r = \dfrac{qQ}{2 \times \pi \times \epsilon_0 \times m \times v^2}$

(ii) $r = \dfrac{qQ}{2 \times \pi \times \epsilon_0 \times m \times v^2}$

$= \dfrac{1\cdot60 \times 10^{-19} \times (29 \times 1\cdot6 \times 10^{-19})}{2 \times \pi \times 8\cdot85 \times 10^{-12} \times 1\cdot673 \times 10^{-27} \times (6\cdot0 \times 10^6)^2}$

$r = 2\cdot2 \times 10^{-13}$ m

(iii) Strong (nuclear) force.

(e) Reverse direction of magnetic field.
Reduce strength of magnetic field.

8. (a) $q \times V = \frac{1}{2} \times m \times v^2$
$1\cdot60 \times 10^{-19} \times 2400 = \frac{1}{2} \times 9\cdot11 \times 10^{-31} \times v^2$
$v^2 = 8\cdot43 \times 10^{14}$
$v = 2\cdot90 \times 10^7$ ms^{-1}

(b) (i) Constant force / acceleration / electric force / electric field in **vertical direction** / **upwards** / **downwards**
Constant motion / speed in horizontal direction

(ii) No unbalanced force / field / attraction acting on electron.

(c) (i) $E = \dfrac{V}{d}$

$= 100/0\cdot01$
$= 10^4$ Vm^{-1}
$F = QE$
$= 1\cdot60 \times 10^{-19} \times 100/0\cdot01$
$= 1\cdot60 \times 10^{-15}$ N
$a = \dfrac{F}{m}$

$= \dfrac{1\cdot60 \times 10^{-15}}{9\cdot11 \times 10^{-31}}$

$= 1\cdot76 \times 10^{-15}$ ms^{-2}

(ii) $t = \dfrac{s_H}{v_H}$
$= 5\cdot17 \times 10^{-10}$ s
$v_v = u_v + a_v \times t$
$= 0 + 1\cdot76 \times 10^{15} \times 5\cdot17 \times 10^{-10}$
$v_v = 9\cdot10 \times 10^5$ ms^{-1}

(d) Length scanned decreases.
v_H increases / greater acceleration.
Shorter time between plates
or vertical speed is less on leaving the plates.

9. (a) $f = 2\cdot4$ Hz

(b) $\lambda = 0\cdot5$ m
$v = f\lambda$
$= 2\cdot4 \times 0\cdot5$
$= 1\cdot2$ m s^{-1}

(c) $y = 4\cdot3 \times 10^{-2} \sin 2\pi (2\cdot4t + 2\cdot0x)$

10. (a) Fringes produced by **interference** of light reflected from **top** and **bottom** surfaces of the film.
Different thicknesses / wavelengths / positions / angles affect colours.

(b) For non-reflecting coating,
Optical P.D. $= \lambda/2$
for **destructive** interference
Optical P.D. $= 2nd$
$2nd = \lambda/2$
$d = \dfrac{\lambda}{4 \times n}$

(c) $2 \times n \times d = (1\frac{1}{2}) \times \lambda$
$2 \times 1\cdot3 \times d = 1\cdot5 \times 7\cdot80 \times 10^{-7}$
$d = 4\cdot50 \times 10^{-7}$ m

(d) (i) $E_k = \frac{1}{2} \times m \times v^2$
$4\cdot12 \times 10^{-21} = \frac{1}{2} \times 1\cdot43 \times 10^{-25} \times v^2$
$v = \sqrt{\dfrac{2 \times 4\cdot12 \times 10^{-21}}{1\cdot43 \times 10^{-25}}}$
$v = 240$ ms^{-1}
$p_{ru} = m \times v$
$= 1\cdot43 \times 10^{-25} \times 240$
$p_{ru} = 3\cdot43 \times 10^{-23}$ kgms^{-1}

(ii) $p_{ph} = \dfrac{h}{\lambda}$
$= \dfrac{6\cdot63 \times 10^{-34}}{7\cdot80 \times 10^{-7}}$
$p_{ph} = 8\cdot50 \times 10^{-28}$ kgms^{-1}

(iii) $N \times p_{ph} = p_{ru}$
$N \times 8\cdot50 \times 10^{-28} = 3\cdot43 \times 10^{-23}$
$N = 4\cdot04 \times 10^4$ photons

11. (a) unpolarised light
$\Rightarrow$ (Electric field vector) oscillates or vibrates in all **planes**
polarised
$\Rightarrow$ (Electric field vector) oscillates or vibrates in **one plane**

(b) (i) $\theta = 0°$
$\theta = 180°$

(ii) Measure I and θ for at least 5 different values of θ.
Plot a graph of I against $\cos^2\theta$.
Graph produced is a straight line through the origin.

ADVANCED HIGHER PHYSICS 2011

1. (a) (i)
$$E = mc^2$$
$$2 \cdot 08 \times 10^{-10} = m \times (3 \cdot 0 \times 10^8)^2$$
$$m = \frac{2 \cdot 08 \times 10^{-10}}{9 \cdot 0 \times 10^{16}}$$
$$m = 2 \cdot 3 \times 10^{-27} \text{ kg}$$

(ii)
$$m = m_o \times \left(\frac{1}{\sqrt{1 - \frac{v^2}{c^2}}} \right)$$
$$2 \cdot 3 \times 10^{-27} = 1 \cdot 673 \times 10^{-27} \times \left(\frac{1}{\sqrt{1 - \frac{v^2}{(3 \cdot 0 \times 10^8)^2}}} \right)$$
$$v = 2 \cdot 1 \times 10^8 \text{ m s}^{-1}$$

(b) (i)
$$E_k = \tfrac{1}{2} m v^2$$
$$3 \cdot 15 \times 10^{-21} = 0 \cdot 5 \times 1 \cdot 675 \times 10^{-27} \times v^2$$
$$v^2 = 3 \cdot 76 \times 10^6$$
$$v = 1 \cdot 94 \times 10^3 \ (m \ s^{-1})$$
$$p = m v$$
$$= 1 \cdot 675 \times 10^{-27} \times 1 \cdot 94 \times 10^3$$
$$= 3 \cdot 25 \times 10^{-24} \text{ kg m s}^{-1} - \text{given}$$

(ii)
$$p = \frac{h}{\lambda}$$
$$3 \cdot 25 \times 10^{-24} = \frac{6 \cdot 63 \times 10^{-34}}{\lambda}$$
$$\lambda = \frac{6 \cdot 63 \times 10^{-34}}{3 \cdot 25 \times 10^{-24}}$$
$$\lambda = 2 \cdot 04 \times 10^{-10} \text{ m}$$

(c) (i) Strong (nuclear) (force)
(ii) 10^{-14} m
(iii) Quark

2. (a) (i)
$$I_{rod} = 1/3 \ m \ l^2$$
$$= 1/3 \times 0.040 \times 0.30^2$$
$$= 1 \cdot 2 \times 10^{-3} \text{ kg m}^2$$

(ii)
$$I_{wheel} = (5 \times I_{rod}) + m_{(rim)} r^2$$
$$= (5 \times 1.2 \times 10^{-3}) + (0.24 \times 0.30^2)$$
$$= 6 \times 10^{-3} + 0.0216$$
$$= 0.0276$$
$$= 0.028 \text{ kg m}^2 - \text{given}$$

(b) (i)
$$v = \omega r$$
$$19.2 = \omega \times 0.30$$
$$\omega = \frac{19.2}{0.30}$$
$$\omega = 64 \text{ rad s}^{-1}$$

(ii) A
$$\omega = \omega_o + \alpha t$$
$$0 = 64 + \alpha \times 6.7$$
$$\alpha = -\frac{64}{6.7}$$
$$\alpha = -9.6 \text{ rad s}^{-2}$$

(ii) B
$$\tau = I \times \alpha$$
$$= 0.028 \times (-) 9.6$$
$$= (-) 0.27 \text{ Nm}$$

3. (a)
$$\omega = \frac{2\pi}{T}$$
$$= \frac{2 \times 3.14}{5.6 \times 24 \times 60 \times 60}$$
$$= 1.3 \times 10^{-5} \text{ rad s}^{-1} - \text{given}$$

(b)
$$F_C = F_G$$
$$M_2 \omega^2 r = \frac{GM_1 M_2}{r^2}$$
$$2.0 \times 10^{30} \times (1.3 \times 10^{-5})^2 \times 3.6 \times 10^{10}$$
$$= \frac{6.67 \times 10^{-11} \times 2.0 \times 10^{30} \times M_1}{(3.6 \times 10^{10})^2}$$
$$M_1 = 1.2 \times 10^{32} \text{ kg}$$

(c) (i)
$$E_P = -\frac{GM_1 M_2}{r^2}$$
$$= -\frac{6.67 \times 10^{-11} \times 2.0 \times 10^{30} \times 1.2 \times 10^{32}}{3.6 \times 10^{10}}$$
$$= -4.4 \times 10^{41} \text{ J} - \text{given}$$

(ii)
$$v = r \omega$$
$$= 3.6 \times 10^{10} \times 1.3 \times 10^{-5}$$
$$= 4.68 \times 10^5$$
$$E_k = \tfrac{1}{2} mv^2$$
$$= \tfrac{1}{2} \times 2.0 \times 10^{30} \times (4.68 \times 10^5)^2$$
$$= 2.2 \times 10^{41} \text{ J}$$

(iii)
$$E_{total} = E_K + E_P$$
$$E_{total} = 2.2 \times 10^{41} + (-4.4 \times 10^{41})$$
$$= -2.2 \times 10^{41} \text{ J}$$

(d) Frequency increases or blue shift when star approaches.
Frequency decreases or red shift when star recedes.

4. (a)
$$y = A \sin \omega t$$
$$\omega = \frac{2 \times \pi}{T}$$
$$= \frac{2 \times 3.14}{5.7}$$
$$= 1.1$$
$$y = 2.9 \sin 1.1 \ t$$

(b)
$$a = -\omega^2 y$$
$$= -1.1^2 \times (\pm) 2.9$$
$$= (\pm) 3.5 \text{ m s}^{-2}$$

(c) F_{max} occurs at $y = \pm 2.9$ m

(d)
$$E_k = \tfrac{1}{2} m \omega^2 (A^2 - y^2)$$
$$= \tfrac{1}{2} \times 4.0 \times 10^4 \times 1.1^2 \times (2.9)^2$$
$$= 2.0 \times 10^5 \text{ J}$$

(e) (i) Period unaffected
(ii) Amplitude is reduced

5. (a) (i) Bring a <u>negative</u> charged rod close to the balloon,
earth (touch) sphere,
remove earth,
remove rod.

or

Touch 2 balloons together, bring charged rod near one,
separate balloons **before removing rod**, identify
which balloon is positive.

(ii) $E = \dfrac{Q}{4\pi\varepsilon_o r^2}$

$= \dfrac{(+)120 \times 10^{-6}}{4\pi \times 8.85 \times 10^{-12} \times (0.35)^2}$

$E = (+) 8.8 \times 10^6$ NC^{-1} or Vm^{-1}

(ii)

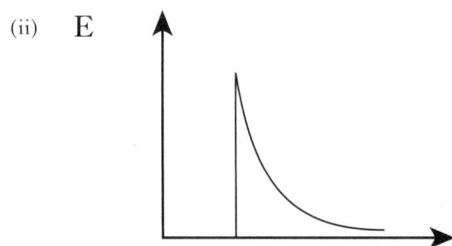

(b) (i) $F = qE$

$E_w = Fd$

$E_w = qV$

$\cancel{q}V = \cancel{q}Ed$

$V = Ed$

$E = \dfrac{V}{d}$

(ii) $V = E \times d$

$V = 7.23 \times 10^4 \times 489$

$V = 3.54 \times 10^7$ V

(iii) $I = \dfrac{Q}{t}$ & $P = IV$

$I = \dfrac{5.0}{348 \times 10^{-6}}$

$I = 14367.8$ A

$P = 14367.8 \times 3.54 \times 10^7$

$P = 5.1 \times 10^{11}$ W

(c)

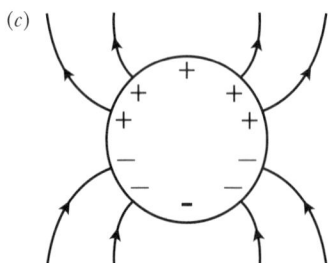

6. (a) (i) Increasing/changing current leads to increasing/changing
magnetic field causes a back emf.

(ii) $E = -\dfrac{dI}{dt} L$

$-12.0 = -\dfrac{dI}{dt} 0.6$

$\dfrac{dI}{dt} = 20$

$\dfrac{dI}{dt} = 20$As^{-1}

(iii) An inductor has an inductance of 1 Henry if an emf of
1 volt is induced when the current changes at a rate of
1 As^{-1}.

(iv) Generates a **large** (back) emf

due to **<u>rapid</u>** change
or **<u>collapse</u>** in **B-field**.

(v) $V = IR$

$12.0 = I \times 28$

$I = \dfrac{12.0}{28}$

$I = 0.43$A

(b) 99kmh^{-1} = $\dfrac{99000}{3600} = 27.5$ms^{-1}

$v^2 = u^2 + 2as$

$0^2 = 27.5^2 + 2 \times -1.0 \times s$

$0 = 756.25 - 2s$

$s = \dfrac{756.25}{2} = 378$m

Yes before the signal.

(c) (i) Wavelength, $\lambda = \dfrac{v}{f_s}$

$\lambda_{obs} = \dfrac{v}{f_s} - \dfrac{v_s}{f_s}$

The observed frequency, $f_{obs} = \dfrac{v}{\lambda_{obs}} = \dfrac{v}{\dfrac{1}{f_s}(v - v_s)}$

(ii) A $f_{obs} = f_s\left(\dfrac{v}{v - v_s}\right)$

$f_{obs} = 294\left(\dfrac{340}{340 - 28.0}\right)$

$f_{obs} = 320$ Hz

B $f_{obs} = f_s\left(\dfrac{v}{v + v_s}\right)$

$f_{obs} = 294\left(\dfrac{340}{340 + 28.0}\right)$

$f_{obs} = 272$ Hz

7. (a) (i) Towards Y.

Cancellation of B-field between the wires
or opposite magnetic fields caused by each wire cause
attraction.

(ii) $\dfrac{F}{L} = \dfrac{\mu_0 I_1 I_2}{2\pi r}$

$\dfrac{F}{L} = \dfrac{4\pi \times 10^{-7} \times 4.7 \times 4.7}{2\pi r \times 360 \times 10^{-3}}$

$\dfrac{F}{L} = 1.2 \times 10^{-5} \, \text{Nm}^{-1}$

(b) (i) $F = \dfrac{0.0058 + 0.0061 + 0.0063 + 0.0057 + 0.0058 + 0.0062}{6}$

$F = 0.0060 \text{N}$

$F = B1l$

$6.0 \times 10^{-3} = B \times 1.98 \times 0.054$

$B = \dfrac{6.0 \times 10^{-3}}{1.98 \times 0.054}$

$B = 0.056 \text{T}$

(ii) Scale Reading uncertainty (SRU)

± 1 digit $\Rightarrow \pm 0.0001 \text{N}$

Random uncertainty (RU)

$= \left(\dfrac{\text{max} - \text{min}}{n} \right)$

$= \left(\dfrac{0.0063 - 0.0057}{6} \right) = 0.0001 \text{N}$

$\Delta F = \sqrt{\text{SRU}^2 + \text{RU}^2 + \text{calibration uncert}^2}$

$\Delta F = \sqrt{0.0001^2 + 0.0001^2 + 0.00005^2}$

$= 1.5 \times 10^{-4} \text{N}$

(iii) $\dfrac{\Delta B}{B} = \sqrt{\left(\dfrac{\Delta F}{F}\right)^2 + \left(\dfrac{\Delta I}{I}\right)^2 + \left(\dfrac{\Delta l}{l}\right)^2}$

$\dfrac{\Delta B}{B} = \sqrt{\left(\dfrac{1.5 \times 10^{-4}}{0.0060}\right)^2 + \left(\dfrac{0.02}{1.98}\right)^2 + \left(\dfrac{0.0005}{0.054}\right)^2}$

$\dfrac{\Delta B}{B} = \sqrt{8.12 \times 10^{-4}}$

$\dfrac{\Delta B}{B} = 0.029$

$\Delta B = \pm 0.002 \text{T}$

8. (a) $F = BIl (\sin \theta)$ **or** $F = BIl$

but $I = \dfrac{q}{t}$

$v = \dfrac{l}{t}$

substituting to get

$F = B \dfrac{q}{t} vt$

(b) $F = \dfrac{mv^2}{r} = Bqv$

$v = \dfrac{Bqr}{m}$

$v = \dfrac{3 \cdot 6 \times 10^{-3} \times 1 \cdot 6 \times 10^{-19} \times 2.8 \times 10^{-3}}{9 \cdot 11 \times 10^{-31}}$

$v = \dfrac{1 \cdot 6128 \times 10^{-31}}{9 \cdot 11 \times 10^{-31}}$

$v = 1.77 \times 10^6$

$v = v_{total} \times \sin\theta$

$\dfrac{1.77 \times 10^6}{2.0 \times 10^6} = \sin\theta$

$\theta = 62°$

(c) Radius decreases.
Pitch increases.

9. (a) Slits/gaps in horizontal and vertical direction.
Explanation of interference pattern.

(b) $\lambda = \dfrac{d\Delta x}{D}$

$4.88 \times 10^{-7} = \dfrac{d \times 8.0 \times 10^{-3}}{3.6}$

$d = 2.2 \times 10^{-4} \text{m}$

(c) (i) B

Larger λ gives larger Δx.

(ii) D

As horizontal d increases horizontal Δx decreases.

As vertical d decreases vertical Δx increases.

10. (a) A stationary wave is formed by the **interference** between waves, travelling in **opposite** directions or **reflecting** from the end supports.

(b) (i) $T = mg = 4.02 \times 9.8 = 39N$

$f = \dfrac{1}{2l} \sqrt{\dfrac{T}{\mu}}$

$f = \dfrac{1}{2 \times 0.780} \sqrt{\dfrac{39}{1.92 \times 10^{-4}}}$

$f = 290 \text{Hz}$

Note is D.

(ii) $2 \times$ answer to **10.** (b) (i)
$f = 2 \times 290 = 580 \text{ Hz}$

Hey! I've done it

¡BrightRED PUBLISHING

© 2011 SQA/Bright Red Publishing Ltd, All Rights Reserved
Published by Bright Red Publishing Ltd, 6 Stafford Street, Edinburgh, EH3 7AU
Tel: 0131 220 5804, Fax: 0131 220 6710, enquiries: sales@brightredpublishing.co.uk,
www.brightredpublishing.co.uk

Official SQA answers to 978-1-84948-232-5
2007-2011